Grand Canyon For Sale

Grand Canyon For Sale

*Public Lands versus Private Interests in
the Era of Climate Change*

Stephen Nash

UNIVERSITY OF CALIFORNIA PRESS

University of California Press, one of the most
distinguished university presses in the United States,
enriches lives around the world by advancing scholarship
in the humanities, social sciences, and natural sciences. Its
activities are supported by the UC Press Foundation and
by philanthropic contributions from individuals and
institutions. For more information, visit www.ucpress.edu.

University of California Press
Oakland, California

Library of Congress Cataloging-in-Publication Data

Names: Nash, Steve, 1947- author.
Title: Grand Canyon for sale : public lands versus private
    interests in the era of climate change / Stephen Nash.
Description: Oakland, California : University of
    California Press, [2017] | Includes bibliographical
    references and index.
Identifiers: LCCN 2016058132 (print) | LCCN 2017000660
    (ebook) | ISBN 9780520291478 (cloth : alk. paper) |
    ISBN 9780520965249 (ebook)
Subjects: LCSH: Grand Canyon (Ariz.)—Environmental
    conditions. | Public lands—Environmental aspects—
    United States.
Classification: LCC GE155.A6 N37 2017 (print) | LCC GE155.A6
    (ebook) | DDC 333.78/30973—dc23
LC record available at https://lccn.loc.gov/2016058132

Manufactured in the United States of America

26 25 24 23 22 21 20 19 18 17
10 9 8 7 6 5 4 3 2 1

For America's government- and university-based natural scientists and their allies

CONTENTS

MAP INSERT FOLLOWS PAGE 22

ACKNOWLEDGMENTS

Abundant thanks to all those whose names appear in the Sources pages near the end of this book. Many of them patiently endured long strings of questions that were at times repetitive and ill-formed.

Some on that list were especially generous with time, information, and insights. They merit special thanks: Debra Donahue, professor, College of Law, University of Wyoming; Martha Hahn, former chief of science administration and resource management, Grand Canyon National Park; David Uberuaga, former superintendent, Grand Canyon National Park; Roger Clark, program director, Grand Canyon Trust; Peter Lattin, former contractor for the Bureau of Land Management; Lori Makarick, vegetation program manager, Grand Canyon National Park; Ken McMullen, former program manager, Grand Canyon National Park, retired; James Nations, former director, Center for Park Research, National Parks Conservation Association; Bruce Sheaffer, comptroller, National Park Service; and Beverly Stephens, special assistant to the comptroller, National Park Service.

I am indebted, too, to Chris Zganjar of The Nature Conservancy and Katharine Hayhoe and Sharmistha Swain of Texas Tech for their projections of regional climate change and national parks; the graphics guru Scott Zillmer of Mapping Specialists (now at National Geographic); and the journalists Vickery Eckhoff and Paul Rogers for their revelations about who really controls federal grazing land.

Biggest thanks to my wife, Linda, who brought an endless supply of patience and affectionate support to this project, along with her indispensable editorial eye and ear. I'm an exceedingly fortunate author, and husband.

CHAPTER ONE

Brink

I hope that you will not . . . mar the wonderful grandeur, the
sublimity, the great loneliness and beauty of the Canyon. Leave
it as it is. You can not improve on it. The ages have been at
work on it.   Theodore Roosevelt

This faint old path isn't on the brochure map, but it leads to a fine perch just the same. Moving past the car choreography and selfie poses at the popular Desert View area near the eastern border of Grand Canyon National Park, I find my way on a late afternoon.

Crumbling pavers end in a trace that weaves through rabbitbrush and juniper and over to a suitable rock, right on the abyss. No glance out there yet. I don't want to risk vertigo until I'm settled. Then, with a beer and a bag of salt peanuts, I can drift out over two billion years of geology, a hundred centuries of human striving, and a timeless void.

Anywhere you pause along the countless miles of edge brings dizzying contrast. The infinitesimal meets the cosmic, as a cliff swallow careens against far-off rock and sky. The immediate—check your footing on that limestone grit, there's a long fall pending—opens abruptly onto silent eons of cycle and revision. Another contrast: under a longer gaze the wild and timeless look of this panorama bears the lasting marks of recent human activity. They are the destinations of this book.

If you were to make your way along this edge with me, for example, your line of sight would sweep out across the immense Venus, Apollo, and Jupiter Temples; the striated vanes of the Cardenas and Escalante

Buttes; and the Tanner side canyon with its own side canyons and their side canyons falling away below in a ragged regression. Across the main Canyon, out on the high North Rim, a horizon of cool evergreens beckons, albeit through ten miles of haze.

Somewhere over there an odd tribe of hybrid beefalos romps and ruminates. With luck we might spot a condor, one of a small number that have been arduously and expensively reintroduced to the Canyon region after the species was shot and poisoned to near-extinction.

The sun eases into a quickening descent. Shadows overtake the nearer depths and climb toward me. From somewhere near come the startling whumps of an air-tour helicopter. Far below and barely visible, the bone-cold blue curves of a segment of the Colorado River seem to define the very bottom of the world, though the name "Colorado" derives from what once was the river's warm, rust-red flow. With binoculars you can make out a green fringe of tamarisk trees along one bank.

Unkar Creek shows there too. It has ferried its cargo of silt and rock off the North Rim and down to the river over unknown millennia, whenever the rains come. The result, a rounded delta, has pushed a wide bend into the Colorado, and added some fine rapids. Humans have farmed that little fan of dirt off and on for perhaps ten thousand years.

Tree-ring data show that the climate, which dictates the flows of the creek, has varied widely over time and directs the presence or absence of constellations of plants, animals, and ancestral humans. Just now Canyon life unfolds within the most severe drought the region has seen since the 1500s. It is reckoned by scientists to be either a signal of human-driven climate change or a mild preview of what it will be like.

The frigid clarity of the once warm and silty Colorado, new since the advent of the Glen Canyon Dam upstream, has all but extinguished native fish species. The prolific tamarisk trees, arrivistes from central Asia, have severely disrupted native plant and animal communities along the river's banks. Then there's the dirty haze from smokestacks and tailpipes, and the hybrid beefalo herd that is overgrazing and eroding fragile rangeland, where such animals (even native buffalo) have never roamed. All are evidence of lapses in our stewardship of this globally revered sanctuary. So is the ceaseless staccato of the air-tour choppers, and the fact that the condors are still being poisoned.

As we head into its second century, few would disagree that we want the park system to fulfill its mandate to preserve nature. "The core element of the national parks is that they are in the perpetuity business," as Gary Machlis, science adviser to the director of the Park Service, told me. "The irony is that our mission is to preserve things in perpetuity, and we do it on an annual budget and a four-year presidential cycle." The natural systems of the parks, he said, represent an island of stability—as long as we protect them and plan well for their future.

As it happens, the view from here on a South Rim rock also takes in other national treasures. To the north and south are the 2,500 square miles of the Kaibab National Forest, contiguous with the park. Off to the east is the 400-square-mile Vermilion Cliffs National Monument, even more remote and far less frequently visited. They are all of a piece with hundreds of thousands of square miles of other national parks, forests, deserts, grasslands, and wildlife refuges.

Conservation is a major part of the official purpose of those other public lands, at least on paper. We've come to recognize, little by little, that they are part of the foundations of our own survival. They could prove to be an ark for what's left of our natural heritage—one that may remain buoyant, if we're supremely vigilant and profoundly lucky.

From a high orbit the Grand Canyon resembles an exquisitely detailed origami. To guess its future well, we have to unfold it outward into full context and see the entire portfolio of public lands that surround it. We've carved up the landscape administratively among the Park Service, the Forest Service, the Bureau of Land Management, the Fish and Wildlife Service, and several other agencies, and we oversee it with greater and lesser levels of protection. Within the horizons of the South Rim, though, you can feel its unity—something like the blanket of muscle strands, blood paths, and neural nets that make up your own body. What remains of their integration has become indispensable to the biological survival of these lands as a whole—more so today than before the Europeans, or the Paiutes, or their predecessor humans arrived.

Federal public lands total more than a million square miles—28 percent of the national dirt. Outside of the national parks, most of us pay those expanses of rock, range and forest little heed, but they're the source of billions in annual federal revenue and easily the nation's single most valuable hard asset. It affords us hiking, hunting, fishing, wildlife habitat, flood and erosion control, and a buffer against climate change, as well as timber, mineral, and oil, coal, and gas deposits. Twenty percent of our clean water is provided by federal forests and grasslands.

Interesting, then, that the U.S. House of Representatives voted to value federal lands as worthless, as the Donald Trump administration took office. This eases their transfer into state or private hands without compensation to the nation. Once begun, each step in that process would be irreversible. A national hunters' and anglers' group responded: " . . . the concept of public lands was born of a desire to remove the shackles of a stifling European system in which only the wealthy or royalty could enjoy the outdoors. We the people own these mountains and forests, rivers and plains. Nothing could be more American." The president then ordered a "review" of established protections for about 20,000 square miles of federal lands, calling them "an abuse." That move spikes the odds for more mines, wells, and commercial development—and far less protection for natural areas.

So public lands are up for grabs. Recent events—fire and insurrection, for example—have brought them sporadically to media attention. They have led Sean Hannity of Fox News to the rhetorical question, "By the way—why do they [the feds] own all that land?" The online hipster news source *VICE* brings a different sensibility but the same question when it tries to delve into the strange novelty of it all, that "the federal government owns the majority of the land in eleven western states, which is, to be fair, a shit-ton of land." Our political and legal history explains how this came to be, but not why it should continue.

The most compelling answer to that question is outside both history and politics, and Grand Canyon is a fine place to begin looking for it.

I've talked with dozens of people who have skin in that continent-sized game, the management of public lands for which Americans hold the deeds of trust. As we now contemplate the disintegration of this estate, it's useful to hear from scientists, administrators, ranchers and developers, environmentalists, and power-plant operators. Through them, we can hope to see more clearly the condition of the land itself, and especially the natural systems it supports. We want to know, don't we, whether we're getting ripped off?

President Theodore Roosevelt, sometimes clairvoyant, advised an admiring crowd on these matters during a May afternoon in 1903. He had decided to visit the national preserve that his leadership would almost single-handedly create in the face of bitter local political opposition. Roosevelt said, "What you can do is to keep it for your children, your children's children, and for all who come after you, as one of the great sights which every American if he can travel at all should see." He implored us (which could not have been easy, given his disposition), "to do one thing in connection with it in your own interest and in the interest of the country, to keep this great wonder of nature as it now is.... I hope that you will not ... mar the wonderful grandeur, the sublimity, the great loneliness and beauty of the Canyon. Leave it as it is. You can not improve on it. The ages have been at work on it, and man can only mar it."

A few years later Congress passed the Organic Act of 1916, which established the National Park Service and defined its mission: "to conserve the scenery and the natural and historic objects and the wild life therein and to provide for the enjoyment of the same in such manner and by such means as will leave them unimpaired for the enjoyment of future generations."

Legal scholars have richly documented the subsequent history of the act and its vaguely contradictory instructions. Were the parks going to be more or less a collective Teddy Roosevelt wilderness or Central Park, or something in between? Was the original intent to emphasize "enjoyment" by U.S. citizens? That calls to mind some comfortable sightseeing or a well-ordered recreational setting and "pleasuring ground," to use a phrase of those times. The "enjoyment" must not harm the nature of the parks, the act states, but the word does appear there twice, after all.

Or should we focus instead on the sinew, the rarest quality of what the writer Wallace Stegner (or Lord James Bryce, or someone) first called "America's best idea?" That obligation is "to conserve … unimpaired" the living organisms, the wildlife, the scenery.

The conflict between the two motives is sharpest right there at the word "unimpaired." We've been deciding what that really means all during the century since the Park Service was founded, and we will continue to. Original intent and original equivocations aside, this is political history as much as law. It follows the pragmatism of the great Supreme Court justice of that era, Oliver Wendell Holmes, who once wrote that "the secret root from which the law draws all the juices of life" is in reality "considerations of what is expedient for the community concerned."

Despite the power of precedent and the illusion of stability conferred by legal language, in other words, the meaning of a law adapts to society's priorities over time. The Organic Act made its promise about leaving nature and wildlife unimpaired. Grand Canyon and the rest of the national park system give abundant evidence of where the promise has been kept, and broken, in the century since then.

When the Park Service drew up an ambitious post-World War II development agenda called Mission 66, it pushed for more roads, more building construction, and more commercial infrastructure for the national parks, all to increase the numbers of tourists and the comforts of their travel. That dominant idea had to make some room for a competing urgency that first arose during the 1970s, however.

A national environmental consciousness was taking root, along with a new federal framework for environmental protection—the National Environmental Quality Act, the Clean Air and Clean Water Acts, and the Environmental Protection Agency. By then the parks' natural systems had already become visibly fragile, increasingly degraded. A new program of science research was launched to assess their condition and prognosis.

These days Grand Canyon scientists transform the acres of spreadsheet data they collect—about bighorn sheep habitat, let's say, or the distribution of cactus species or summer rainfall—into visualizations that are often like sets of maps. These are far better adapted to human comprehension, just as a roadmap is much easier to figure out than a string of numbers that might be used to describe a route.

Decks of these data pictures, or infographics, can be superimposed, one on another. You could see how a certain bird or insect species occurs more frequently at higher elevations, amid particular kinds of vegetation, in specific winter temperature ranges, or near springs and seeps. The science that underlies these sometimes strikingly beautiful graphics can unlock puzzles about how we should manage the land.

There are no resource-management overlays, though, that incorporate some other powerful but less tangible factors that can dominate natural systems. We need to, but cannot, map out the distribution of human political power within a natural setting, how it operates to extend the life chances of certain fish species or the amount of pollution in Canyon air and water. You can't easily visualize the flow of money that enables environmental decay in Arizona. It would be difficult to contrive an overlay to show whether scientific recommendations for managing endangered species are followed, fudged, or dismissed out of hand because of pressure from lobbyists or commercial interests or from a kind of willed ignorance that rejects science.

Too bad, because those influences matter deeply in any realistic picture of the Grand Canyon region's future, that of Great Smoky Mountains National Park, Alaska's Tongass National Forest, or any other federal lands. Instead of a fine set of maps and overlays, we will have to portray political factors by other means.

It's getting chilly out here as the sunset fades. It is fleeing west, past Las Vegas, which for now at least is safely on the far edge of dusk. And I'm thinking, all well and good about how public lands are imperiled, but come on: *Grand Canyon for Sale.* Is that preposterous? "Grand Canyon will always be there!" an acquaintance out east, where I live, recently assured me. I knew exactly what he meant. I've been offered the same warrant by many others, including a top administrator in the Park Service. But in an all-important sense, they and their catchphrase are mistaken.

Grand Canyon comes to mind first as its defining image: a nearly eternal horizon of pillars and walls. A sunken cathedral of rock. All that part of the scenery will indeed, within casual calculation anyway, always be there. But despite the barrenness that a calendar photo may suggest, the park supports a diverse, fragile living realm. That's because it is protected, of course, and because its six thousand feet of abrupt descent yields a broad range of natural communities. Boreal forest on the high North Rim that is more characteristic of Canada transitions down through several life zones to low, scorching desert similar to that of Mexico, at the level of the Colorado.

This jumblestack of habitats shelters 91 mammal species—bears, rodents, bighorn sheep, coyotes, ringtail cats, skunks, raccoons, bobcats, foxes, and cougars among them—57 different reptiles and amphibians, 373 birds, 8,480 kinds of insects, and 17 fish. Twenty-five plant and animal species are formally recognized as extinct, extirpated (gone from the Canyon), endangered, threatened, or "species of concern" for some other reason, and the list is no doubt incomplete.

Frontline, longtime conservation campaigners inside and outside the government can be impatient with bright ideas about the work of protecting that heritage. They've heard them before, and, really, they might easily conclude, so what? They don't have time to indulge thumbsucker fantasies about best-case scenarios. They have to fight hard for what can really be accomplished, given the hand they've been dealt. After all, much of the nation thinks the federal budget is busted. And even if you've concluded that that's a hysterical hoax, our current political landscape is gridlocked at nearly every intersection. Who's going to step up for a panoramic fight over public lands?

That's an easy call, in a way—a process of elimination. Government officials we've hired to protect public lands show memorable courage at times, but they are also obliged to be loyal to their chain of command. We can't ask them to "die on every hill," in the constant battles they face against economic and political interests with priorities other than conservation.

So federal administrators can really be only as brave and good as we allow them to be. Absent reliable and highly visible public support they give ground, and compromise, and as time passes our civic inattention—the fond, false hope that someone else is minding the store—accelerates the hazard to the whole public landscape. What if the store is steadily pilfered or has caught fire? Environmental groups, too, have to work largely within the circumscribed options of politics, the "art of the possible," that we Americans create for them.

The reason to take a look at the current state of the Grand Canyon—the foreground example for this book—and our other parks and public lands is not to pretend that more new ideas are all we need. It's to see what needs to be done. "There is this constituency that's uninformed, that takes their national parks for granted, that they'll always be there and they will always be cool, and they just don't really know how threatened they are," as former Grand Canyon superintendent David Uberuaga told me. "They don't realize how powerful their voices are with their congressional delegation and that there's a need to be raising hell."

The centenary of the Park Service has just passed, along with some well-deserved national self-congratulations. Perhaps this would be a discreet time to say that the parks' natural systems are, in the estimation of many scientists, falling apart. In that view all public lands need long-term life support, beginning as soon as we can pull it together. We're on a precipice, both politically and biologically. It's a good time to visit those scientists and their research, and take a look around before deciding our next moves.

My working theory is that if you look hard enough, you can measure whether we've lost far too much ground over the years in that well-rehearsed, real-world pragmatism of political compromise. It may be as if we're running up an escalator, and we can see progress if we focus on each step. But we have to look around, instead, to see if the reality is that we're moving backward and down.

If you're a reader like me, you're wondering by now what my agenda is—as in, political agenda. Here are my biases, then: I'm part of a small-business-oriented family, I'm a journalist, and I teach. I have strong faith in the historical, productive ingenuity of a market economy. I have also drawn continual inspiration from the commitment, wisdom, and tireless labor of scientists and administrators in public service whom I've met over the last thirty years of writing about their work.

As it happens, though, that work unfolds within an electoral system that can foster astounding waste and self-dealing. Indeed, the only source of disappointment in public life that rivals government sometimes is the blind rapacity of unregulated free enterprise. We can vote out misbehavers in government, at least—except when the electoral system itself is nearly owned outright by just a few.

Naturally, I wonder what's on your mind too. My assumptions are that you may or may not know much about them, but you have a firm enough attachment to the United States' national parks, forests, monuments, wildlife refuges, and other public lands, and the natural life within them, to read about their future.

That's my concern too. An incisive painting by the artist Robert McCauley shows a big stoic brown bear, standing erect and ready to speak behind a cluster of press-conference microphones. If only it could! For now and ever, though, we humans have to see how nature fares, and speak on its behalf.

If you question any of the fact-assertions in these pages—and you really should—their sources are usually evident in the text. If not, please consult the detailed endnotes. Some of what's here may strike you as quarrelsome or overconciliatory, but it isn't fiction.

I guess it might instead be offered as a secular prayer—for the salvation of Grand Canyon, the national parks, and the much wider public-lands legacy. Long ago I attended a family dinner at which the patriarch gave this invocation: "Dear Lord, help us figure it out." That's all we really have, to start with. And in that spirit of both humility and hubris, let us carry the inquiry forward (see Map 2).

CHAPTER TWO

Alien Abductions

Things get by you.
Joseph Cavey,
federal inspector

Grand Canyon's 5.5 million annual visitors penetrate its 1,900-square-mile expanse along different roads. Most, 93 percent, arrive on the teeming South Rim. Only six in a hundred take the long, sinuous road to the more remote North Rim. It is only 10 line-of-sight miles across from the South Rim, but 24 miles of strenuous hiking by trail or 212 miles and more than four hours away by car.

Another path into the park begins at an easily missed turnoff from Highway 389, about three hours northeast of Las Vegas. It is 61 miles of washboard—a potholed, spike-rocked, tire-killer dirt road that ends at a place called Tuweep on the western edge of the park. And that's the *least* challenging way in to this lonely, lightly visited area.

There's no drinking water, electricity, gas, food, land line, cell-phone service, or shelter at Tuweep. It is just a ranger station with—only sometimes—a ranger, and the enormities of the Canyon and its surrounding plateaus. Fewer than three in a thousand park visitors come here each year, despite the lure of the tiny, crowd-free campground and its compelling scenery: a sheer, three-thousand-foot drop to the river.

These surroundings might strike you as a quintessentially southwestern landscape, hard to imagine anywhere else on the planet—

timeless, limitless, overpowering, wild, native. Surprising to find, then, that even the forbidding sheet bedrock and sparse soils in this remote part of the Canyon are a showcase stocked with novelties: imported and unwelcome souvenirs from Asia, Europe, and Africa.

One hike from Tuweep is a treacherous scramble route of two miles down to the Colorado. You arrive at Lava Falls—a half-submerged remains of ancient lava-flow dams, and the most treacherous rapid on the river. It routinely flips rafts, rafters, and cargo. Just upstream among the rocks along the river are dense groves of the Eurasian tamarisk tree, sometimes called salt cedar. It is a good introduction to a phenomenon rapidly unfolding here and over the rest of the continent too.

The ecologist Charles Elton, more than half a century ago, made the word "explosion" a technical term for "terrific dislocations in nature" caused by the recent waves of invasive species such as tamarisk. "We are seeing huge changes in the natural population balance of the world," he wrote. "We must make no mistake. We are seeing one of the great historical convulsions in the world's flora and fauna."

We're usually aware of this invasion only as seemingly unrelated bits of news: fire ants, killer bees, kudzu, "starry sky" beetles, "stinky trees," stink bugs, pythons, snakehead fish, or zebra mussels. But the dots connect. This is a hydra-headed epidemic that goes by various names—biological pollution, exotic invasions, alien species introductions.

The home ranges of most insects, diseases, plants, and animals were comparatively stable until we humans began moving them. In natural conditions their ranges grow and contract over decades or centuries in response to climate or other changes in the environment, but they rarely jump thousands of miles or across high mountain ranges or oceans on their own. The faster our global travel and trade accelerates, the more species move to new places.

Humans arrived in the Southwest at least as early as twelve thousand years ago, doubtless bringing along a few new species that were favored food or medicinal plants. The arrival of Europeans quickened the conveyor belt. The seeds of one of the first nonnative invasive species brought in from Europe, for example, was probably a stowaway among ballast rocks in the holds of ships in the early 1600s, dumped on the shores of Jamestown, Virginia. Now this highly toxic plant pest can be found coast to coast: its debarkation point has survived in its nickname, "Jimson weed."

Today the appearance of alien species that we bring into new areas is so common worldwide that it far surpasses natural rates of change. One count of free-living alien plants, land vertebrates, insects, spiders, fish, molluscs, and plant diseases in the United States found that about fifty thousand kinds have found their way, through our activities, into agricultural and natural areas, including Grand Canyon.

It's hard to predict how a newly introduced species will behave. As a federal study puts it, "the movement of plants, animals, and microbes beyond their natural range is much like a game of biological roulette." Or Russian roulette. The climate, or native predators and competitors, quickly kill off most new arrivals. But for others their new home is a paradise. They inhabit an ecosystem with few or no defenses. Tamarisk trees, for example, coevolved on their native continent with competitors and insects that kept their numbers in check. None have had time to arise here. Instead tamarisk populations leap, and out-compete natives like cottonwood trees and willows that many wildlife species depend on.

Invasives often thrive on disturbed landscapes. They can disperse quickly, tolerate a broad range of climates, and have easier success in places where their native competitors have weakened or disappeared because of grazing, acid rain, ozone pollution, erosion, logging, mining, and drilling. Research confirms that many invasives will prosper within the torn-up ecology of global climate change.

This conversation may strike you as reminiscent of our acrid national political debate over human immigration. There's no connection. The problem with alien invasive species is not that they're foreign. It's that they overwhelm native plants and animals, weakening the whole natural web. Biological diversity—the variety of species on a landscape—is one rough index of the health of an ecosystem. But the biological diversity that alien invasive species bring is brief and destabilizing. It leads away from diversity in short order, to the sameness to be seen at Lava Falls—a blanket of uniformity that many native plants and animals cannot adapt to.

In Grand Canyon this is abundantly evident in the case of cheatgrass, brought to the United States in the mid-1800s. Cheatgrass is now found throughout the Canyon. It invades areas after a fire, pushes out native perennials, and completely changes the natural fire cycle. In this way the whole complex of native plants, insects, and animals is disrupted or killed off, the basic rhythms of their ecosystem shifted out of gear.

Hundreds of alien invasives like tamarisk, cheatgrass, kudzu, wild pigs, giant cane toads, snakehead fish, and balsam woolly adelgid insects are a factor in nearly half of the cases in which threatened or endangered species are declining across the United States. A federal report warns that as invasives overtake the national parks, the change "may be so severe that they will eliminate the very characteristics for which the parks were originally established —as refuges for America's native ecosystems and species." Near Yosemite the destructive New Zealand mud snail may soon join yellow starthistle, the algae called didymo (or sometimes "rock snot"), and a long list of other unwelcome imports. At Washington State's Olympic National Park, it's Himalayan blackberry, English ivy, English holly, and Japanese knotweed.

Yellowstone has two hundred alien species, including the New Zealand mud snail and an introduced lake trout that threatens the native cutthroat trout, as well as the European whirling disease parasite, which kills off salmon and other fish species. At Hawai'i Volcanoes National Park, natural resource manager Rhonda Loh told me, the number of invasives has increased steadily during her thirty-year career, and the trend is likely to continue. "I don't see anything stopping that," she said. Her park has a catalog of invasives, from nonnative feral pigs to mosquitoes, which did not exist in the islands before Europeans came. Argentine fire ants have arrived too, "but they've been eclipsed by Little fire ants," she said.

"You think it's bad, and then something else comes in." Ohia trees, an iconic rainforest species, are being decimated by an imported sudden-death fungus.

The Everglades is now overrun by melaleuca trees, ambrosia beetles, Burmese pythons, lionfish, sharp-toothed tegu lizards and dozens of other destructive non-natives. Eastern forests, including those in Shenandoah and Great Smoky Mountains National Parks, have lost nearly all their populations of several tree species, including hemlock, butternut, chestnut, and Fraser fir, to introduced insects and fungi. Ash and dogwood are being decimated by Chinese borers and an imported fungus, while Asian pest trees such as Paulownia and Tree-of-heaven move into even the most remote wilderness areas.

Invasives and noxious weeds are already the dominant vegetation on an estimated fifty-five thousand square miles of public lands, and they spread at a rate of some seven square miles per day, according to the Bureau of Land Management. They "threaten soil productivity, water quality and quantity, native plant communities, wildlife habitat, wilderness values, recreational opportunities, and livestock forage, and are detrimental to the agriculture and commerce of the U.S." Trying to shovel our way out from under that blizzard of bad news, we might well be asking ourselves a couple of questions about how this all happened and what we're doing to stop the importation of still more invasive alien species before they too set up housekeeping.

Native ecosystems, including those in national parks, resemble islands on the map, but in some ways that's misleading. In terms of our defenses, the real borders of the Grand Canyon ecosystem extend to Chicago, Los Angeles, Miami, Seattle, and beyond—any port of entry where invasive species are streaming in from other continents. Then the winds, the roads, and the Colorado River carry many of the invaders into the park.

Today there are nearly two hundred species of nonnative plants in the Canyon—about 10 percent of all the flora here. Some were introduced deliberately to the region long ago as feed for livestock, as ornamentals, or for erosion control. More recent arrivals come in, or threaten to, because our inspection processes at coastal ports are porous.

There are nonnative animals to be reckoned with too, as they displace natives and forage and breed destructively: elk, wild pigs, and burros. Usually the obstacle to getting rid of them is the tenacity of the invasives themselves in defending their adopted turf. At other times the problem is turf battles waged by some among our own species.

A hybrid of cattle and bison often called "beefalo" have taken up residence on the Park's North Rim. Bison are not native here, though they might have been rare visitors before European settlement. Cattle are not native anywhere on this continent. The beefalos' foster parent is the Arizona Game and Fish Department, but the herd dates from the Teddy Roosevelt era. A private experiment to provide game animals for hunters, they were sold to the state in the 1920s. Responding to hunting pressure, the beefalo moved first to the national forest, whose administrators acceded to their presence on behalf of hunters, and eventually to the park, whose administrators have not.

The herd has quadrupled in size, to about four hundred animals. Attempts to work with the state game department to remove them have been met with indifference until recently. "They eat a lot," a park resource manager told me, "and this is not an environment that has evolved for grazing. They've pretty much stripped the vegetation. It's a large concentration of big animals in a small area." Negotiations continue, she said.

About eighty of Grand Canyon's alien plant species are known to spread rapidly, and, in fact, about half of Grand Canyon's overall acreage has been altered by aliens. Park lands are at risk from species with colorful and faintly sinister names, such as Dalmatian toadflax, rush skeletonweed, puncturevine, houndstongue, foxtail barley, and bull-thistle, Scotch thistle, musk-thistle.

The work of holding back this biological tide, or at least managing it, falls to people like Lori Makarick, Grand Canyon's invasive plant specialist. Part of her work is helping supervise a battalion of volunteers who commit some 17,500 hours each year to hacking, pulling, and poisoning invasives in the park. Think of it as a struggle that pits one determined species—us—against a multitude of others.

Each of the volunteers who labor on Canyon invasive species crews is probably unique in experience and motivation, and Hannah Heinrichs is no exception. A scientist and entrepreneur, she lives in Germany but travels frequently to the United States. Visiting Flagstaff a few years back, she was urged by friends not to miss the Grand Canyon. Only slightly interested, she made time for the trip on a whim. She has returned three dozen times since then, often as a volunteer. It's hard labor.

"Usually in the morning the whole vegetation crew and all volunteers meet at the bullpen, where each of the group leaders grabs his or her share of volunteers. If you are assigned to invasive removal you have to make sure that you have the right gear with you or on the truck," she says. "In most cases this means safety vest, gloves, geopick, clippers and plastic bags. In special cases it means safety glasses, helmet, and a Pulaski—a kind of axe-adze tool. Early in the season, when the invasives are not flowering and have no seed heads, it is easier. Later it means extra caution, because one false move can spread tens of thousands of seeds. Usually you clip the seed heads first and bag them and then remove the rest of the plant."

The volunteers usually have to walk from plant to plant, bend over at every one and remove it, carrying tools, bags full of removed plants and personal gear over long miles of trails and bushwhacks. "You want to get the plant out with all its roots, so there will be no regrowth," Heinrichs says. "If the soil is soft and the plant is shallow-rooted, just pull. But, hey, that's wishful thinking at the Grand Canyon. Most invasives are deep-rooted and the soil is, more than often, hard as rock."

Most of the invasives "don't like being pulled," she said, and they are well defended. Bull-thistle, for example, has spikes that penetrate even thick leather gloves. You have to learn a certain technique for each species to remove it permanently without hurting yourself. It has been suggested that what's needed, here and on public lands across the United States, are tens of thousands more Hannah Heinrichses, well paid for their difficult, tenacious labor. That investment would be a bargain. Prevention is an even better one.

Tamarisk is one of the aliens that has preoccupied Makarick. A native of Kazakhstan and Iran, according to genetic tests, it first showed up in the 1930s. Tens of thousands of tamarisks line the Colorado all through the Canyon. Millions more have altered the banks and the nature of countless rivers and streams in all but thirteen states. The heaviest and most damaging infestations extend from the state of Washington south into Mexico, and from east Texas west to the Pacific.

Tamarisk is efficient at pulling water out of the landscape—up to three hundred gallons a day—with a tap root that can descend ninety feet into rock and soil, robbing native plants of water even as its dense foliage grabs available sunlight. It can easily grow three feet each year. Tamarisk needles have high salt content too, and when it sheds them the salt permeates the soil and fends off native competitors like cottonwood and willow. Finally, that long tap root makes even very young tamarisk seedlings devilishly hard to yank out.

At the park it spread and became dominant after Glen Canyon Dam was built in the 1960s. The controlled flow of water from the dam has fostered the buildup of a new ribbon of sediment along the banks of the river, where dense stands of tamarisk now thrive. Before the dam that sandy, silty habitat would have been scoured out each year during spring floods.

Now tamarisk has colonized the river corridor shoreline, and spring winds blow its seeds up into the side canyons. "When we've looked into those side canyons and tributaries that still have their natural flow," Makarick says, "we realized that we are almost at a now-or-never point in some ways. So about ten years ago, we started looking at how feasible it would be to remove tamarisk from such places."

Native tribes such as the Hualapai have partnered with the Park Service in this arduous and endless campaign, supplying field crew leaders, planners, and pullers. A battalion of anti-invasives troops, paid and volunteer, returns to the side canyons every two or three years to hunt tamarisk. A quarter of a million have been removed.

"In all these side canyons where you've got tamarisk growing in abundance there's really not much else," Makarick says. "None of the native forbs, grasses, or other riparian species that might have been there. It's all just completely altered. But once the tamarisks have been removed, water sometimes emerges again on the surface. You start to see cattails, and horsetails, and other marsh and wetland species begin to return that had been there before the tamarisks invaded."

With companions I once left the Grandview Trail and Horseshoe Mesa to ease down along the walls of the narrowly beautiful Cottonwood Creek side canyon, where Makarick's crews had cleared tamarisk only the year before. We kept finding vigorous new seedlings that stubbornly resisted attempts to pull them, all along the watercourse.

Makarick sometimes has bad dreams about tamarisk and the other Canyon invasives, like Scotch broom or Rush skeletonweed, that she tilts against. In one dream she walks through a beautiful field of grasses, "and then I just start to have this tremendous itchy reaction, and I realize that cheatgrass seeds are embedded in my legs and I can't get them out."

Can't get them out. That sums up the status of most other invasives too. Our history of attempts at eradicating invasives sometimes looks like variations on the theme of the old Donald Duck cartoon in which he pursues a pest—a fly—with increasing vexation until he finally grabs a shotgun and destroys his whole house. The fly buzzes around vigorously in the wreckage.

So at times we've tried a disastrous course of poisons, like the enduring insecticide Mirex, which was used to kill Argentine fire ants until it began to show up in the food chain, in human breast milk, and in the Great Lakes. We've also tried "biological control," which introduces another alien species in the hope that it will attack the first one. You can guess how this can go terribly wrong, and it often has. These days long and thorough testing is used to try to make sure the new alien preys only on the invasive we're trying to get rid of. That helps reduce the odds of a second disaster, but there is still risk.

A new player on the Grand Canyon scene is one of these biocontrol introductions, the tamarisk beetle. Once released elsewhere, it spread here unexpectedly and inadvertently and illustrates how biocontrol can morph into out of control. The beetle defoliates tamarisk trees—something to cheer about—sometimes often enough to kill them. So far, it has not been seen to harm native plants.

In general, however—unless they're intercepted very early—when invasives are here, they're here to stay. With supreme effort and expense they may be managed to minimize the worst consequences. Scotch broom, for example, dominates parts of the inner Canyon at lower elevations, and its tenacity is also impressive. What would it take to rid the Canyon of this particular blight? "Oh gosh, I don't think it's possible," Makarick told me. "It would take an alternate reality. You would need a massive supply of native seed, and I don't think anyone in the Southwest has that. You would really need to seed native species for years and try to tip that balance. I just don't think it's possible."

A group of scientists once abandoned the typically dry language of research to write, "when the outrageous economic and ecological costs of the wanton spread of existing exotics and continued entry of new ones become common knowledge, it is inevitable that there will be a public outcry for action to mitigate the potentially dire consequences." But that was decades ago. "Dire" has long since arrived and the outcry, so far, is muted, despite billions of dollars in annual losses to agriculture and uncounted losses on natural landscapes. Grand Canyon, and every other national park, forest, and grassland bear the real cost of the damage from invasives, and it is orders of magnitude greater than the cost of effective inspection measures. The most recent estimate of the *annual* cost of invasive species in the United States is $120 billion.

Australia and New Zealand are also battling invasives, but they have instituted aggressive inspections of ships and shipping containers at their borders, and quarantines for any imported plant material. U.S. ports of entry have chosen against effective protection. For now the pace of new introductions of invasives is undeniable evidence that our defenses are not serious compared to the scale of our self-inflicted damage.

"Things get by you," a recently retired federal inspector, Joseph Cavey, told me. He takes pride in the hard work of his agency but acknowledged, "We don't have security at all the piers in the country. Customs tries to handle that, and we try to handle our part of it, but it just doesn't happen." By the end of his career at the Agricultural Plant Health Inspection Service, or APHIS, the lead federal agency for controlling alien invasives, he was a branch chief of the group that figures out the identity of plants, insects, animals, and diseases that turn up at national borders.

"In the most common scenario, and it probably occurs a hundred times daily, inspectors intercept a pest—insects, plant diseases, plants—in cargo, or they might find it when they take something away from a passenger who is bringing in food that they're not allowed to bring in. It's no secret and we freely admit that we can't get all this stuff. There's just too many shipments to look at them all. Only a very small percentage is actually inspected."

The agency's playbook is tightly restricted, in any case, by politics and budgets. Its chief mission is to protect agriculture, not the environment. Bugs that don't threaten crops or farm animals sometimes get a free pass to enter the United States, even if they might invade forests and the rest of the nonfarm landscape and start chewing. "If we were to intercept a praying mantis that's exotic to the United States—say it just

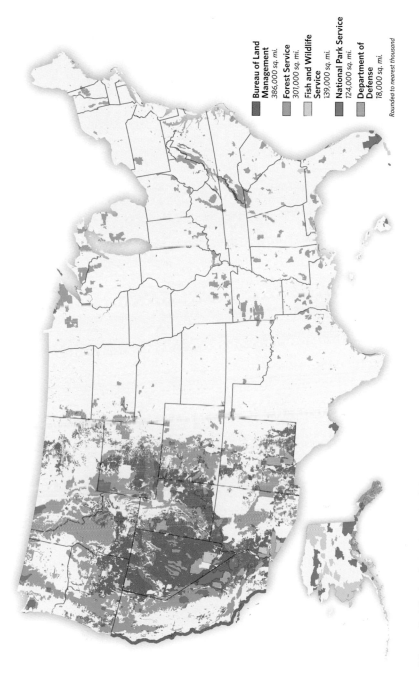

Map 1. Public lands in the United States.

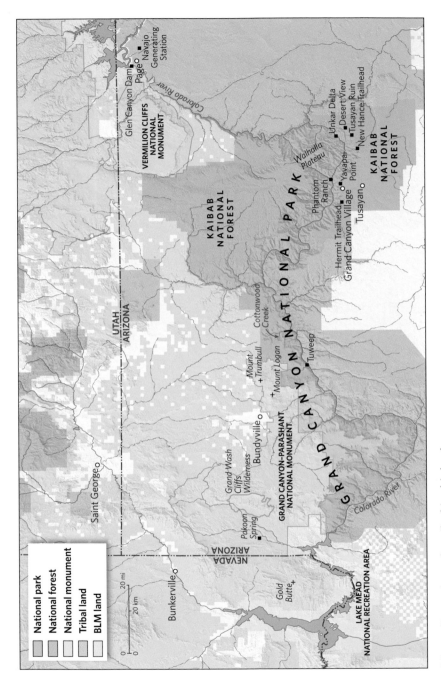

Map 2. Some places described in this book.

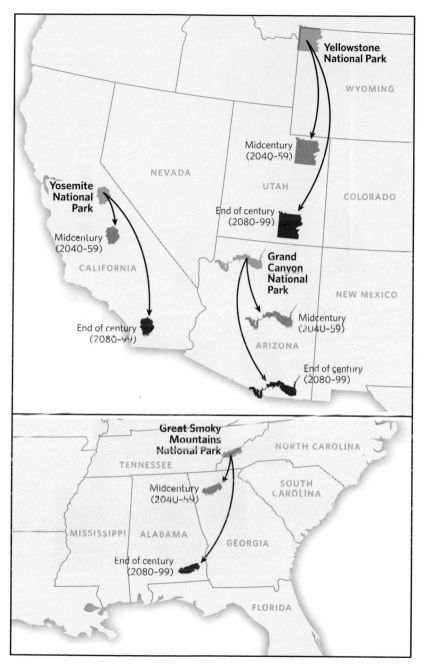

Map 3. Shifting climate, shifting nature.
In a "business as usual" scenario, in which greenhouse-gas emissions continue to accumulate along current trend lines, national parks will effectively move to hotter, more southerly latitudes as the Earth warms. Chris Zganjar, The Nature Conservancy.

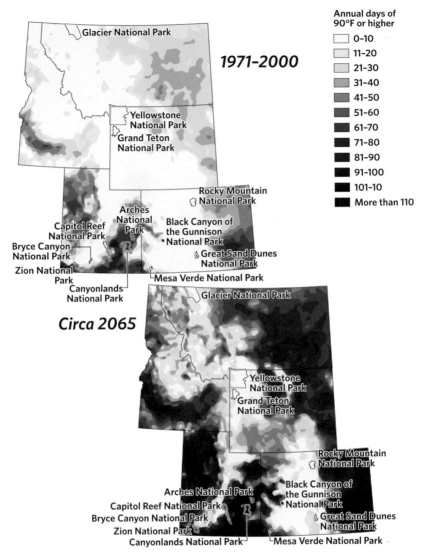

Map 4. How much hotter will national parks be?

On average Grand Canyon National Park's South Rim saw 61 days a year of temperatures 90 degrees Fahrenheit and above from 1971 to 2000. That number is projected to increase, on average, to 110 days a year by around 2065—if greenhouse-gas emissions continue to gather over the planet on a "business as usual" path and global temperatures surge to more than 5.4 degrees over historical averages. This map shows similar comparisons for national parks to the north of Grand Canyon. Map by Katharine Hayhoe and Sharmistha Swain, Texas Tech University.

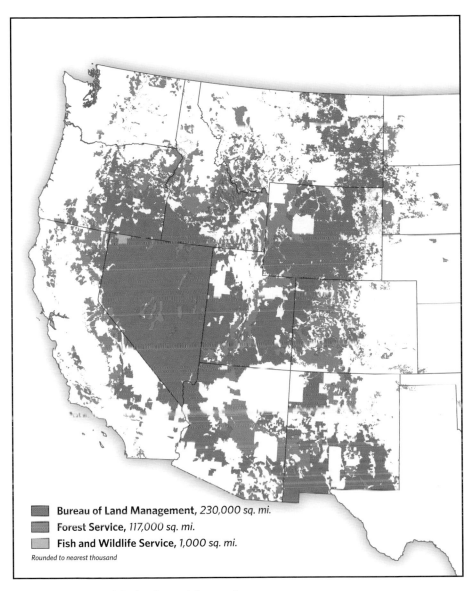

Bureau of Land Management, *230,000 sq. mi.*
Forest Service, *117,000 sq. mi.*
Fish and Wildlife Service, *1,000 sq. mi.*
Rounded to nearest thousand

Map 5. Western public lands used for grazing.

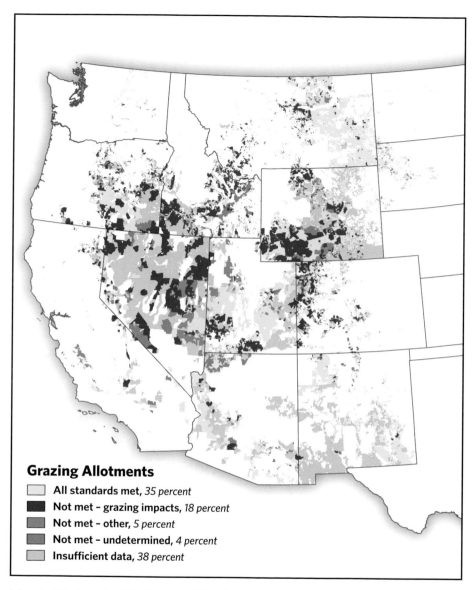

Grazing Allotments

- ☐ All standards met, *35 percent*
- ■ Not met – grazing impacts, *18 percent*
- ▨ Not met – other, *5 percent*
- ▨ Not met – undetermined, *4 percent*
- ▨ Insufficient data, *38 percent*

Map 6. The health of the Bureau of Land Management's grazed land.
BLM range-health records are inconsistent and incomplete. This map shows an
approximation, using BLM sampling data, of where grazing damage has occurred. See
chapter 8 and endnotes for limitations on the map's accuracy.

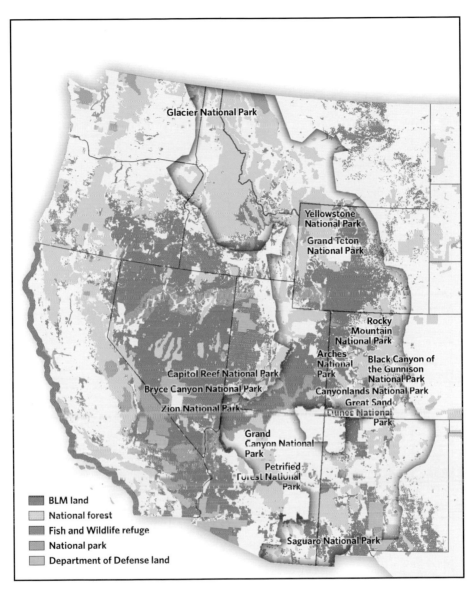

Map 7. Linking public lands for survival.
To aid in their survival, plants and animals will need protected lands through which they can migrate as climate change proceeds. This is one outline, proposed by the nonprofit Wildlands Network, for how public lands might be consolidated.

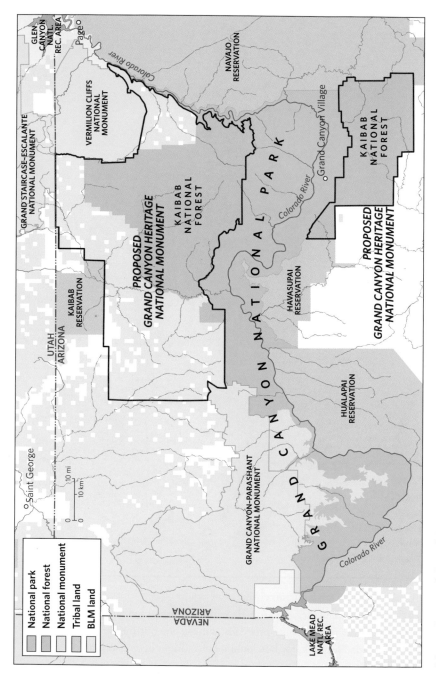

Map 8. Proposed Grand Canyon Heritage National Monument.

wandered into a container shipment of tiles, and it got here—we would not be able to take action for that," Cavey told me. The mantis could perhaps wreak havoc on the natural landscape, but it is a predator, not a plant pest, so no action could be taken. This rule is part of international agreements designed to prevent countries from gaining an unfair trade advantage. The worry is that those countries could restrict imports by faking concern about alien invasives.

"Yeah, you're betraying the ecology," Cavey said. Potentially harmful invaders could be interdicted, but "we are not empowered to do that." Praying mantises are actually not so common, but other predators come every day. "Ground beetles, carabid beetles—gee, there's so many—certain wasps that are strictly insectivorous, for example. And those are not our charge," he said. Nobody else's charge either. He and others have campaigned, with limited success, to expand the watch lists to include at least some known environmental threats.

Cavey, a pragmatist, did not advocate policy changes, citing real world constraints on budgets and lack of political support. I pressed him to imagine what might occur if those factors were off the table and keeping invasives out was the overriding priority instead. He said, "If you look hard enough, you can find something alive in most shipments.... And listen, what we export to other countries, same story. This could be completely stopped, if that country were to break the trade conventions and say, 'You know what? Any new living thing, if it were to become established here, could upset our ecology,' and made that the utmost concern rather than trade." In the United States, "if that were the priority, if that came down from Congress and through the Executive Office, then there is definitely more we could do. You bet." But if you want that, you'll need much more funding too, he added.

A federal survey of several agencies found that a shortage of money is a common theme in fighting invasives. Their comments highlight "the impact of inadequate resources on the ability of agencies to respond to new infestations"; "inadequate resources or attention to the problem";

and "agencies' inability to fund accelerated research on emerging threats has limited the availability of effective control methods."

No one needs reminding that federal budgets are a point of contention among our leaders, but that is not the only possible source of the funds needed for an effective inspection system. We might do well to consider instead the "polluter pays" principle, long a part of environmental regulation. In this case international traders who covet the U.S. marketplace are also the polluters who bring in invasive species. The United States imported about $2.3 trillion in goods during the most recently tallied year. Tax that traffic at just a tenth of a cent on the dollar, and you'd have a war chest of $2.3 billion a year to fund inspections to keep new invasives out and to cope with the ones already here. Why has that not occurred?

Other polluter-pays strategies have long been proposed: insurance requirements, bonding, inspection fees, and civil and criminal penalties and fines. The real problem, then, is political. Traders don't want to pay fees to keep invasives at bay. More careful inspections and tighter restrictions may slow down international trade, and that brings political opposition. "It happens all the time," Cavey said, with a figurative shrug. "Industry is going to push back. They're going to try to get the best deal they can get." He offered the example of importers of landscape nursery plants, who balked at standardizing inspection routines so that insects and diseases could be intercepted more effectively. The longer plants are in transit, the more they weaken and die, so the shipper's game is to speed things up, not slow them down. If the agency stands its ground, it invites political interference—calls from members of Congress on behalf of their corporate constituents.

So the agency may attend to those affected by its mission rather than the mission itself. Those people have money and durable influence to bring to bear, while advocates for natural systems are fewer, poorer, easier to ignore. APHIS has even been heard on occasion to call the shippers its "customers." This warns of the possibility of "regulatory capture," which turns government agencies from watchdogs into lapdogs. The federal government itself defines regulatory capture as occurring "when an agency becomes dominated by the industry it is charged with regulating and acts in ways that benefit the industry rather than the public interest."

Cavey's wing of APHIS deals only with plant pests; a different part of the agency handles imports that might affect agricultural livestock. Threats to the environment from imported animals boil up through a maze of other federal agencies that are also susceptible of political pressure. More than 2,200 different kinds of alien wild animals were legally imported into the United States in one recent four-year period, according to the records of the U.S. Fish and Wildlife Service—many of them destined for the pet trade. And 302 of those birds, fish, reptiles, mammals, and other species—more than one in ten—were predicted by scientists to be invasive or harmful in the United States.

Consider the saga of nine different species of alien constrictor snakes that some pet fanciers like to keep. They include the Burmese python, which is already an epidemic in Everglades National Park—they are big enough to eat alligators—and Africa's largest snake, the mean-tempered, twenty-foot-long giant rock python. The senior herpetologist at the Florida Museum of Natural History, presumably a snake fan, called this species "just one vicious animal."

In 2006 a Florida water district petitioned the U.S. Fish and Wildlife Service to prevent the importation of those nine snake species. A decade later nothing had been done. "Unfortunately, when it came to weighing the economic interests of these few breeders against the enormous economic and ecological damage these snakes can cause ... a small but vocal sector of the pet industry concerned with importing and breeding these dangerous exotic snakes seems to have put a stranglehold on a sensible rule," Dr. Bruce Stein of the National Wildlife Federation said, years into the comatose process. Finally, a rule against importing these snakes was imposed. Hundreds of other alien animal species are still legal imports.

Even at a remote place like Tuweep, the roll call is startling: the tumbleweedy Russian thistle is so widespread over this huge landscape that

Park Service crews who visit once a year eradicate it only from the little campground. Stork's bill, native to the Mediterranean basin, is a common resident too. Scotch thistle, whose origins are in western Asia and Europe, grows here, and so does horehound, which came from North Africa, Asia, and Europe. That's also where the ubiquitous cheatgrass comes from, but no one tries to control it in the park anymore either. London rocket thistle has shown up around Tuweep, and it may be just a matter of time before Sahara mustard—a North African arrival that can grow four feet tall and smothers and robs moisture from native plants—arrives from Glen Canyon National Recreation Area upstream, where it is now abundant.

As things stand, the burden of risk of new invasives, and the expense of dealing with what's here already, are borne by the general public, the continental landscape, and native plants and animals rather than by the polluters. Compared to the tens of billions of dollars of annual damage from invasives, a rigorous system to keep them out would be dirt cheap. It could reasonably be borne by the shippers and their clients.

It's not just about money. It is also that Congress would have to stand up to special interests that profit from this long, destructive snooze. Congress has regularly considered bills that could fix our broken system for keeping out invasives. Meanwhile, they continue to arrive. You could say it's like flicking lighted matches into dry tinder and hoping that nothing much will happen.

CHAPTER THREE

Landscapes in Motion

All the evidence coming in is that it is going to be much worse than we've expected.

David Breshears, ecologist

The North Rim of the Grand Canyon is about eight thousand feet above sea level, so snow arrives early and stays long. An average winter sees twelve feet of it. The few roads disappear, unplowed, for the duration. The old gray lodge shuts down in mid-October and doesn't come back to life until seven frigid months have passed.

One sector of the North Rim, the Walhalla Plateau, is lower and surrounded on three sides by the Canyon, so air currents rise from the bottom to bring warm days a bit earlier in spring. Out there along Cape Royal Road, the infrequent traffic mostly whizzes past a small archaeological site called Walhalla Glades. If they could step out from the pine grove there and flag us down, though, the people who animated this silent ruin a thousand years ago could paint a vivid picture of climate change and the life of the Canyon region for us.

Until that occurs, their visible legacy in this little clearing is eight rock rectangles, three courses high. The largest one encloses an area ten feet by thirty feet. In the right late-season light, its ragged walls suggest a brief era when these were head-high, roofed storerooms for grain or living spaces with hearths. They were a seasonal home and food source for a few families of Ancestral Puebloan people. Dozens of such structures have been found on the plateau, along with serenely beautiful pottery, tools, and jewelry—and irrigation works and human skeletons that do, indeed, tell a climate story.

Natives of this part of the continent made an enormous leap out of their hunter-gatherer culture to learn the practice of agriculture, with new technology and new food—maize, beans and squash—probably around the year 900. Shifting climatic conditions in a generally dry period sometimes brought more rain to the region, opening up large expanses of newly arable land.

"It was perfect for farming," the archaeologist Douglas Schwartz told me, "and the human population exploded. As ecologists have learned, that always happens when a species moves into a new region. At Grand Canyon the climate was so good that the populations on the South Rim began to increase and spill over into the bottom."

Schwartz, one of the godfathers of Canyon archaeology, mounted the first large-scale excavations here in the 1960s. If you walk to an overlook across the road from the Walhalla Glades, you can make out, at a bend far down along the distant Colorado River, one focus of that work: the Unkar Delta, the largest population center in the Canyon bottom at its peak, circa 1050 to 1150 A.D. A few of its inhabitants somehow migrated up along the near-vertical Canyon walls each summer to the North Rim, to increase the community's hard-pressed food supply. They also found cooler temperatures. They captured snowmelt in irrigation systems and farmed the Walhalla Plateau.

The final chapter for these particular settlements began only a few generations after their arrival. The encouraging climate began to dry out. That diminished the farmed crops, the dozens of wild plants that were also harvested, and the game, especially deer.

"You begin to see evidence of malnutrition, famine, and conflict," Schwartz said. Skeletal remains from the region show the ravages of hunger and of violence. Food was more elaborately protected against roving, starving raiders. During its archaeological surveys, Schwartz's team managed to scale the formidable Sky Island, an isolated tower butte I could see from the Walhalla overlook. They were surprised to find what they concluded were food-storage structures even there. They demanded enormous labor to build, stock, and defend in such a remote and difficult place.

"And pretty soon none of this mattered," Schwartz said, "because everybody was gone." Over a few decades, native populations gave up waiting for the return of a more forgiving climate. They abandoned the whole Canyon region and moved on.

Today, the climate is shifting again. We have begun to calculate the force and hazard of that epochal change. As coal, oil, natural gas, and other fossil fuels are combusted and as fire consumes more of the remaining forests of the planet, carbon atoms are released in carbon dioxide gas. It wafts gradually into the atmosphere and stays there, some of it for centuries and about a quarter of it for as long as a hundred thousand years. That blanket of greenhouse gases traps long-wave radiation that would otherwise pass into space. The added radiation is heating Earth.

"Moderate to extreme drought continues to cover nearly all of Arizona," was the snapshot one agency recently reported. "More than 72 percent of the state is classified with severe drought, while extreme drought covers about 16 percent of the state. Almost all of New Mexico is under moderate to exceptional drought (the highest category), with more than 77 percent of the state designated as severe and over 40 percent designated as extreme drought."

Downstream from Grand Canyon is Lake Mead, behind Hoover Dam. By 2017 it had dropped to its lowest elevation since the lake filled in the 1930s, thanks to a drought that began in the year 2000. Upstream, the levels at Lake Powell mark the lowest fourteen-year period since the Glen Canyon Dam, which created the lake, was completed in 1963.

A recent federal climate assessment tells us what may seem obvious: future heat and drought in the Southwest will ramp up stress on the region's rich diversity of plant and animal species. "Widespread tree death and fires, which already have caused billions of dollars in economic losses, are projected to increase, forcing wholesale changes to forest types, landscapes, and the communities that depend on them," the study says.

It is now possible to sharpen the focus of climate research to a regional scale. Tree-ring studies have allowed scientists to reconstruct a record of rainfall and sometimes heat for the Southwest, back to the year 500. They show that the long current drought is still within the range of natural variation, according to Jeffrey Dean of the University of Arizona. Because climate is the very long average of weather, many climatologists hesitate to attribute one trend or another completely to an overarching phenomenon such as global climate change without at least three decades or so of data.

Instead, they say, what these conditions show us is the kind of climate that will begin to dominate the region as climate change takes hold. With time the odds of more heat and drought ratchet upward. A stretch of hundred-degrees-plus days at Grand Canyon was always possible before global warming. It's just that those long scorchers and even higher temperatures are increasingly likely now and in the very near future that awaits us.

Regional annual average temperatures in the Southwest are projected to rise by a scorching 2.5 to 5.5 degrees Fahrenheit by 2041–70 and by 5.5 to 9.5 degrees by 2070–99 if the world continues with "business as usual" global greenhouse gas emissions. The greatest increases will be in summer and fall. Summertime heat waves will be longer and hotter, and the trend of decreasing wintertime cold air outbreaks is likely to continue.

If, instead, greenhouse gas emissions are very sharply and quickly reduced in the near future, those projected increases are markedly cooler, though still unnerving: 2.5 to 4.5 degrees Fahrenheit (2041–70) and 3.5 to 5.5 degrees (2070–99). Gradually, around the end of the century, the increases are projected to level off, under this work-hard-and-hope scenario. Another line of recent research combines tree-ring records with climate models to construct a drought projection. By around 2050 ecological stress in the region's forests will exceed the severity of the strongest megadroughts since the year 1000 and perhaps further back than that.

I asked the lead researcher for this work, Park Williams of Columbia University's Lamont-Doherty Earth Observatory, whether the findings are as bleak as they seem. No matter what humanity does to rein in greenhouse gas emissions, conditions in the Grand Canyon region will be quite a lot drier than they would otherwise be, he said. After the middle of the century, the projections show a drought that exceeds any period for which we have a record. An average year in the middle of the century will be worse than the worst years of the worst drought in the last thousand years. By the 2080s the best years will be like the worst years of the last thousand, he said.

Under changing conditions, the ability to migrate can be crucial for the survival of humans and natural systems alike, but sometimes change proceeds too rapidly. Southwest forest species may need to move hundreds of feet upslope or hundreds of miles north during the coming century to find climates similar to the ones they have evolved within.

"But the primary conifer species in the Southwest—piñon pine, Ponderosa pine and Douglas fir—are not really good migrants," Park Williams points out. "They generally drop their seeds fairly close to where they are, and those seeds don't get carried far. They are not suited to move quickly in a fast-changing climate."

Wildfire infernos in the Southwest in recent years have created burned-out patches as large as eight hundred square miles. In an area that wide, it could take well over a thousand years for seeds to make their way from the new edge to the center, to recolonize the whole burn.

Many Canyon-region animal species are also threatened by climate disruption, such as the rare Mexican spotted owl, for one of dozens of examples. Its nesting success depends on enough rain to grow plants that in turn support the small animals that owls prey on. Drought-generated wildfires will destroy owl habitat and spread diseases, like the West Nile virus, that target birds.

Bighorn sheep are currently under study at Grand Canyon because of their vulnerability to several kinds of environmental change, including climate. Their small remaining populations everywhere are isolated and disconnected, the chief wildlife biologist for the Park Service, Glenn Plumb, told me. "We end up with small groups here and there, and they may or may not be able to connect up. If those connections fail to occur, you run the risk of genetic isolation. Or if something happens to a group, they might be wiped out, and other sheep would not be able to repopulate the area."

Bighorns are sensitive to habitat conditions. In fact, they are a sensitive species altogether, Plumb said. "We're building roads, cities, infrastructure—and that is one of a triad of intersecting stressors for this species. The second one is invasives—it could be plants or diseases moving into their habitat. It could be other, nonnative, sheep or animals such as mountain goats that we have picked up and moved, as we have in the Black Hills of South Dakota. Then you lay climate change on top of that. This sensitive animal is smack dab in the middle of this intersection. It has to cope with all of it."

As the impacts of climate disruption accelerate, the fate of many such species will depend on their ability to adapt by migrating to higher, cooler, and perhaps wetter elevations or latitudes while still maintaining connections to their already dwindling populations to avoid inbreeding. If these living systems are to survive, we have to find ways to preserve their chances to move freely, far outside the limits of a park or other protected area. In effect, we are going to have to expand the national parks and consolidate, connect and protect all other public lands to both north and south.

Models project, for example, that the gentle, aromatic forest of Gambel's oak, juniper, and piñon pine that graces the South Rim will disappear from there. The first two species may find higher, cooler places to shift toward over time, but piñon pine, now widely distributed across all of northern Arizona, is likely to disappear from the state altogether. The spruce and fir forests of the higher, chillier North Rim will also be vulnerable. They are very likely to give way, over time, to species that can tolerate more heat and drought. "We'll be living in a quite different world in the Southwest. It could be catastrophic, but there is reason for hope, for maintaining some of our forests if we manage them carefully," Williams said.

The future of forests in the Southwest is also very much on the mind of David Breshears, lead scientist at the University of Arizona's Terrestrial Ecology Lab. His research group studies the relationships between ecology, water, and climate. "It's a little bit of an 'Oh, duh.' Trees die faster under hotter temperatures," he told me. "There's a lot of reason to be concerned that plants are not going to be able to migrate with the change of climate."

Trees have an emphatic influence on the climate around them. They absorb billions of tons of the planet's carbon dioxide, storing it until they decompose or burn and release it back into the atmosphere. Older trees pack on wood growth and carbon storage quickly, growing in diameter rather than height, recent research has found. The older the forests the better, and the less logging the better.

Trees have evolved to capture radiation coming in from the sun. They trap heat and control it by evaporating water. Landscapes with trees, then, have the advantages of a heat-capturing, water-emitting blanket that cools the local climate with shade as it locks up carbon, at least for awhile. Large-scale tree death brings a harsher, hotter local climate. That can make it harder for the next generation of trees to get restarted as seedlings.

"So a lot of forests in the West and around the Grand Canyon are going to be at risk of very rapid change. The problem is that the trees can't necessarily keep up with that pace of change," Breshears said. "Once they are lost, many formerly forested landscapes may just stay barren. There is also much higher potential for invasive species like cheatgrass to come in. We don't have good data on species migration rates. We do know that the trees' sensitivity to temperature is going to jump in and play a big role."

With higher temperatures even a shorter drought can kill trees, he warned, adding that "the more information we get on how sensitive these trees are to temperature, the more dire the situation looks." The threat spans different species, elevations, and latitudes. Climate-related die-offs in stands of aspen, Ponderosa pine, piñon pine, and juniper have already been seen.

"I'm not saying that we're going to lose all forests or that trees won't migrate in some conditions in some places, but I am saying that the magnitude of change is grossly underappreciated, even within the scientific community," Breshears said. "Those of us working on tree mortality continue to see more and more results that concern us. There's a difference between 'alarmist' and 'alarm,' right? One is saying there's a problem when there isn't. But in some cases you have a big problem and you need to pull an alarm."

Breshears is aware that the scale of destruction, and its urgency, are not easy to fathom. "All the evidence coming in is that it is going to be much worse than we've expected," he said. "Hopefully, that's a motivation for action. If you haven't stood in the middle of one of these very large forests where there has been this massive mortality, it's hard to appreciate it. A colored map doesn't really clue you in. This is large-scale dramatic change—a very large proportion of trees in a landscape you loved have suddenly died. And that causes a whole cascade of other changes."

With and without the climate factor, the national park system's trend lines for wild species hardly reassure, despite successes here and there. Populations of the resplendent native birds at Everglades National Park have fallen by an estimated 93 percent even since the 1930s, when the park was established. At Sequoia National Park the namesake giant trees are threatened by future drought and heat. At current rates of melting, the glaciers at Glacier National Park will be gone within twenty years. The glaciers at Denali National Park and throughout the central Alaska Range are "thinning and retreating rapidly," according to the Park Service.

At Great Smoky Mountains, our most-visited national park, one in six mammal species may be gone within the next few decades because of climate change, and the losses continue on after that. Fifty-four plant and animal species are already on the threatened/endangered list at Hawai'i Volcanoes National Park, which can afford programs to try to help only five of them. Many more will join the list.

The trend is nearly as grim at many of the other parks. A full 96 percent of National Park Service land and 84 percent of park units are in areas where warming was already underway during the twentieth century. Without connections to our larger publicly owned landscape, the life-rafts that we built for wildlife when we inaugurated the national park system a century ago begin to sink. In a stable climate, we have been able to defend protected areas to a degree, as natural landscapes with closed boundaries. An unstable, shifting climate is now forcing those boundaries open (see Map 3 and Map 4).

If natural migration starts to fail, we will need to assist in moving plants, animals, and ecosystems. "As more evidence is coming in as to how severe this die-off is going to be," Breshears said, "relocating species on a larger scale is coming up more often in conversation among scientists." Assisted migration is a debate, not a plan at this point. But unfragmented landscapes through which migrations can occur are rare. Making humans the intermediary to physically relocate other species to help ensure their survival may seem loopy, in both scale and intent, but consider the alternatives.

University of Virginia forest ecologist Hank Shugart once thought through this challenge, as it might play out in eastern national forests and parks: "Let's say the climate changed tomorrow afternoon. Your smart move would be to head off down to Georgia someplace and get a whole bunch of plants that would grow in our forest in this new climate and hire every high school kid on the planet to plant them," he told me.

"But it would still take a couple of hundred years to develop the new forest," he said—even in ideal circumstances, there's a delay. That means "you're going to end up for human-lifespan time periods with plants that are either going to be dying, or at least not prospering. You'd have to do something to the existing forest to get new species to repopulate, too. How are you going to plant Georgia pine trees in an oak forest? Well, you have to knock down the oak trees, to start off."

Interventions on anything like that scale are on the far side of speculative, so far. "There's no agreement on it at the moment," Mark Anderson, director of conservation science at The Nature Conservancy, told me. "Those of us like me who are pessimistic about climate change think we've got to let go of the purism and work with nature," by relocating some species. "There is a large voice for that at this organization," he said. "There's another group that is still extremely uncomfortable with it. Because the track record of people intentionally monkeying around with introductions and movements is really poor. And we tend to have created more problems than we ever created solutions."

For now, many of our most sensitive public lands—large expanses essential for climate adaptation—are often given over to other purposes instead. Those uses are usually destructive of wild habitat. The inflexibility of current park boundaries is the Achilles' heel for species preservation on public lands, as a Yale University study puts it—and it is the barrier to adapting to rapid climate change. If the Ancestral Puebloans of the Walhalla Glades and the Unkar Delta had been similarly hemmed in when their climate transformed, they would have had no chance to move and survive.

CHAPTER FOUR

Ghost Tour

*As a society, maybe we haven't quite figured out what kind of
relationship we want to have with nature, and how we want to live
with wolves or, in the case of the Southwest, whether we want to.*
    John Vucetich, wildlife ecologist

You don't have to hike very far to see the exhilarating and nearly extinct
condors of the Grand Canyon region. They glide right over the gift
shops and museums at the South Rim during warmer months. The
swift shadow of that formidable nine foot wingspan flickers over clumps
of visitors gazing skyward, mouths open.

These are North America's largest flying birds, and among its most
endangered. Binoculars pull in the image of a beaky face emerging
from a bizarre collar of black feathers, like a fur stole. The eyes, full of
off-kilter mischief, are portals to some far wild realm. Anthropomor-
phizing—attributing human qualities to animals (or plants!)—is something serious people avoid, of course.

But anyway we all do it, I tell myself as I leave the shops behind, to
follow a trail below the rim and look for condors in a more natural setting—the tilting vistas of Plateau Point, about six miles away. If you
had visited there a couple of seasons back, there's a ghost of a chance
that you might have seen Condor 133. A treasured asset for the program
that is trying to return condors to the region, she produced three
chicks, probably in caves in the Redwall limestone layer some three
thousand feet above the Canyon bottom. Those celebrated condor

hatchlings are all too rare—a total of only three arrived in one recent year, and all died or disappeared.

Plan for a longer trek, and a lot more luck, if you hope to glimpse another charismatic near-ghost whose natural range includes the park. The nearest Mexican wolf—it might just as well be called the Grand Canyon wolf—will be some seventy-five miles southwest of here and part of a tiny, embattled remnant, the rarest subspecies of gray wolf in North America. Wolves inhabited the Canyon region until the last of them were trapped, poisoned, or shot about a century ago.

Although they would probably have been quite rare, grizzly bears may have been here, too, because there was a robust Arizona population. The nearest grizzly now is in the Yellowstone region, six hundred miles north. About 2 percent of former grizzly habitat, and 1 percent of its former population, still endure in the lower forty-eight states. Only two in ten die of natural causes. The other eight are killed by humans, despite their protected status as an endangered species. The grizzly's tightly restricted habitat is declining in terms of its carrying capacity because of climate change and invasive species. The way to ease those pressures is to create more protected grizzly habitat, including in Arizona, according to ecologist David Mattson.

Nimble jaguars once stalked these Grand Canyon ledges too—bundles of wary sinew under an orange hide strewn with black rosettes. They can weigh nearly three hundred pounds and they are eleven feet long, much larger and more powerful than mountain lions.

There's no record of the very last of the jaguars here, but among them were a female and her two kittens, discovered and shot in the Canyon, probably in the 1880s. Two others were killed around Grand Canyon Village in 1918 and 1932. So if you were visiting to sense the presence of this rightful resident, well, it's not anywhere around here either. Instead, you'd have to hike south, for about a month. On very rare occasions jaguars cross from Mexico to reside in the southern Arizona borderlands. As of this writing, only one is known of, in the Santa Rita Mountains near Tucson.

From Plateau Point I can see the Colorado flowing greenly through the Canyon bottom far below. It's the habitat of playful, mercurial Southwestern river otters. They won't be down there today either. They are most likely extinct, everywhere. On public lands—even the national parks—dozens of species of animals have been pushed to that edge and sometimes over it.

The conservation biologist William Newmark was the first to focus research on this question, back in the 1980s: are national parks large enough to hold onto their full complement of species in the future? The answer: not really. He found that among mammal species alone, eighty-four that were known to have lived in the big parks he studied in the United States and Canada were missing—"locally extinct." About half that number disappeared after those parks were established. The Grand Canyon, for example, has lost one in five of its mammals.

National parks are surrounded by nonpark landscapes. The larger the parks are, and the more well protected the lands adjacent to them are, the better the chances that any particular species can survive, Newmark found. Otherwise, any random event—drought, fire, disease, the disappearance of prey or forage, inbreeding—can wipe them out.

Newmark is now a research curator at the Utah Museum of Natural History, and his focus has broadened to parks and preserves around the world. "I think there's a real reason to celebrate," he said. "I don't think we should dwell only on the idea that, yeah, parks have lost species over time; therefore, the parks are a failure. They're a fine success, imitated around the world. How much would have survived if we hadn't achieved this?" But we know, he added, that without more effective conservation systems, we will lose many species in the future, inside the parks and beyond them.

The United States' good fortune is that a third of its land is publicly owned—a higher proportion than nearly any other country. "But how are we managing it? Obviously, it's those lands adjacent to the national parks that are critical," Newmark said. "If we really want to conserve species in the future, we have to think seriously about linking those lands to existing parks. If we manage them to allow free movement of species in response to climate change and to effectively enlarge the parks, then I'm confident we could conserve nearly all of the species there now."

Countries like Costa Rica are protecting larger proportions of their national land base than the United States and giving much more serious thought to conserving what's left of the natural world than we are, Newmark told me, even though some are desperately poor. "The U.S. was really on the forefront of conservation, but in some ways, it has lost the leadership that it once held," he said.

Complex and expensive efforts to rescue the prairie dogs or black-footed ferrets, desert tortoises or bighorn sheep on public lands sometimes succeed to a limited degree. But often they struggle or fail outright—populations of the Mojave species of desert tortoise have declined by 32 percent in just ten years—and for strange reasons. Their endangered status has little to do with the tenacity of the animals themselves—they've survived on these landscapes far longer than we have—or with the science that undergirds their recovery. Conservation biologists have devoted careers to lab and field research, canny work-arounds, and creative technology.

Instead, the at-risk species and the habitats that are essential for their survival have been caught up in an opposing current, an enduring political undertow. It now threatens not only a wide range of wild species but our legal claim as U.S. citizens to public lands themselves. "If the political side was cleaned up," a wildlife ecologist who works on endangered species projects told me, "if the state and federal governments weren't acting in some ways as though they're opposed to recovery, the animals would be all right."

The condor project at Grand Canyon, largely paid for by private groups like the Peregrine Fund, is something of an antidote to worries about endangered species—a hope-inspiring success. A viable wild condor population could point the way toward restoring many missing species to the national park system. It is, after all, the place where nature is to be "left unimpaired for future generations." The reality: the condor campaign is in precarious shape.

Thirty years ago a mere 22 condors remained on the planet—only 9 in the wild—and the species was headed to near-certain erasure at the hands of humankind. Those last wild birds were captured in California to begin a captive breeding program. It led to reintroductions there, in Mexico, and in the Grand Canyon region, where 6 birds were released in 1996 to try to establish a wild population after an absence of seven decades.

More than 70 condors soar over Arizona and southern Utah now; 161 fly free in California and Mexico; another 200 or so are in captive breeding centers. Their survival so far is the result of three decades of focused labor by teams of wildlife biologists and their helpers and tens of millions of dollars in public and private funding.

"Some people think of them just as big vultures," the avian ecologist Jeffrey Walters told me, "but when you start talking about intelligence among birds, they are supersmart. They play, and they remember things for long periods of time. It's like you have a bunch of creatures in the ocean and then you have dolphins. Most people would say, well, dolphins are special. Condors are that kind of bird."

Which is why they've made a bit of trouble at times. In Southern California some learned to mill around at a hang-glider launch area. They would take off with the gliders, flying among them, disconcertingly, just for something to do. They've been known to get inside houses or peck shingles off roofs.

In the early days of the reintroduction, condors found their way into the house of a vocal critic of the program and ripped up his bed. "That's just coincidence, of course," Walters supposes. One group of the birds was allowed to become too comfortable around people while captive. Released at Grand Canyon, they began to loiter at the overlooks, bullying children to try to get their food, almost like a gang of teenage ruffians. "We learned something from that experience about how not to raise condors in captivity," he said.

Condor 133 was hatched at the Los Angeles Zoo and released along with five other birds in the Vermilion Cliffs, north of the park, on December 12, 1996. It was a fanfare day with a phalanx of visiting dignitaries watching, because these were the very first condors set free in the Canyon region. By the time of her death, 133 was the last living member of that original group.

Condor 133's continued survival was always in question, just as it is for any other animal in the wild. For all their gregariousness and at least a couple of million years of endurance as a species, condors are vulnerable now. Females do not reach breeding age for five or six years, usually don't produce their first viable young until age eight, and lay only a single egg every other year, if all goes well. They fall prey to coyotes, golden eagles, lightning, rare collisions with cars and power lines, trash ingestion, and unexplained "nest failures."

Their worst enemy, however—the factor that threatens more than all other known causes of death combined—is not a natural one. It is our management of their habitat, even on public lands, that most often sickens and kills these huge, exceedingly rare, and expensively cherished birds. Of condors in this region about seventy survive, twenty-eight have gone missing, seventeen died of unknown causes, eleven were killed by other animals, and thirty died of lead poisoning.

On the thousands of square miles of public lands around here but outside the park, federal and state agencies allow hunters to use lead bullets instead of copper. Lead bullets explode into a sphere of hundreds of tiny lead fragments on impact in a deer or other animal, poisoning the meat anywhere within two feet of the entry point. This is sometimes called the "snowstorm effect."

Condors are scavengers. They feed on the carcasses and the gut piles of field-dressed animals that are left by hunters. If they ingest lead in any form, they can suffer progressive weakness, starvation, and sometimes abrupt death. They may be more susceptible to predators and less capable of breeding. Parts of their digestive system begin to rot. "It's a horrible way to die," a Canyon wildlife biologist told me. "It can be a long, slow death."

Voluntary use of copper bullets is said to be fairly high among hunters in the Canyon area (during hunting season, at least) and to be making some headway in condor habitat in Utah, which includes huge tracts of federal land. There is hope for ongoing hunter-education programs. In any case, condor poisonings are still rife.

But the U.S. Forest Service has stoutly resisted declaring lead ammunition to be illegal on national forests and, along with the Bureau of Land Management, has customarily deferred to state wildlife agencies about managing game animals. This has led federal administrators—not to mention many state wildlife officials—to tell me routinely that animals are the state's business, even on national forests or rangeland, and that's that. Sometimes Arizona even argues that it has jurisdiction within national parks.

The federal government defended its client agency, the Forest Service, during a recent lawsuit over the condors in federal appeals court. Its posture is that banning lead ammunition won't solve the whole problem, which may be true; that the Forest Service has no responsibility to keep toxic lead off public lands, which is curious; and that the agency's failure to do so is not its fault: it's the state's call, and the state allows individual hunters to decide.

The last argument is false. I don't know this because of my own exalted parsing of the law. I know the Forest Service can ban lead ammunition on the Kaibab National Forest on Grand Canyon's North Rim, and the Bureau of Land Management can ban it at the other end of the condors' range in Utah, because their attorney, Allen M. Brabender, of the U.S. Department of Justice's Environment and Natural Resource Division, said so in open court. "Could the Forest Service, if it was so inclined, ban the use of lead ammunition on the national forest?" he was asked by one of the three appellate judges. "The Forest Service does have that authority," Brabender responded. I suppose, then, that we're left to scratch our chins on behalf of a lengthening list of poisoned condors and try to puzzle out the agencies' lack of motivation.

Jeffrey Walters is the lead author of the most recent assessment, by six scientists, of how the whole condor-management program is faring. The report concludes—along with every other evaluation of the condors' prospects—that their recovery as a wild population is completely stymied by the continued use of lead ammunition.

Multiple captures, blood tests, and treatments for lead poisoning are commonplace for Canyon region condors during their lifetimes. Condor 133 was captured for testing on nineteen separate occasions and treated for lead poisoning a dozen times. Sometimes the crop, a digestive organ deep in her throat, was paralyzed and could no longer move food into her stomach. Without extraordinary medical intervention to extract the lead, she would have starved to death.

For these treatments, the crop is opened and cleaned. A feeding tube is inserted into the bird's stomach, and twice a day food and fluids are administered and a medication is injected. It binds with the lead, allowing the kidneys to remove the poison from the blood. The process continues for days, halts for several more days to allow recovery, then resumes. "Everybody in the program wishes they didn't have to do that anymore," Walters said. "There are a lot of bad things about it. Everybody realizes there's a lot of stress on the birds."

Condors can live as long as seventy years. Condor 133 was found dead at only seventeen. Her lead levels then were not much above normal, and the cause of death could not be determined. But this poison causes permanent nerve damage in many species and could plausibly have compromised her longevity, fitness, flight, and ability to ward off predators.

All this means that despite appearances, the noncaptive condors are not really wild, in the sense of self-sustaining. Instead, their persistence is completely reliant on humans. "If they stopped trapping them and treating them for lead poisoning, they would be extinct in the wild again, probably in twenty years. They'd all be gone. Everybody pretty much knows that's the case," Walters said.

What determines the number of wild condors in this region, then, is how much lead ammunition they ingest and how much money their private sponsors are willing to pour into the project. "We can have as many out there as we want, but the more we have, the more it costs," Walters summarizes. More birds mean more staff to capture more poisoned victims in more places and to perform more treatments. Conservation groups have to calculate, "Why should we have more numbers when it's all artificial, and it just means we have to spend more money and raise more every year?"

For conservation biologists the first rule in the recovery of rare species is to solve the problem that led to the decline in the first place. "If you don't fix that, there's no point. And they haven't fixed it," Walters said. "There's no doubt in my mind that condors went extinct in the wild originally because of lead poisoning, and the lead is still out there."

Tragically, lead bullets poison human children who are fed wild game too. Lead ammunition is banned in all of Japan for that reason, among others. Some programs in the United States that passed out game meat to needy families have been suspended because tests showed that a percentage of the meat had dangerous levels of lead. "It's possible that as many kids are getting lead from deer meat as ever got it from lead paints," Walters said. "For children it's really bad. I suspect there are kids out there that have eaten way too much of that growing up, and it has affected them."

"It's almost as if you have to sue the agencies these days to get them to protect rare species," he said, "because they really don't like to stand up to anybody. They're just afraid of backlash from certain elements in the hunting community, and especially from the National Rifle Association. That's the only organization that we contacted that wouldn't talk to us. We talked with a lot of hunt clubs and a lot of people involved in hunting, but the NRA just flat out said, 'We won't talk to you.' Their line is that this is an attack on hunting by environmentalists." The NRA also declined to talk with me about condors and lead bullets.

Ironically, hunters aid the survival of condors because the remains of killed animals they leave are a key food source, so hunters shooting prey is good for condors. "They just need to shoot with copper bullets instead of lead," Walters said.

It is comforting to consider what has befallen the condors as an isolated case. I thought over the story of Grand Canyon's native wolves—trying to gauge the quality of its reassurance—while picking my way along a milder section of the rugged Nankoweap Trail. It was chiseled out of the northern walls of the Canyon in 1881, following an old Native American path, at the direction of Maj. John Wesley Powell and his geologist Charles Walcott. The trail was intended to make it easier for them to understand the geological order of the spectacular rock strata, though "easy" has never described this route.

It is a good place to listen for wolves—or it was, a century ago. The waxing moon on a night like this one might have provoked howls that would make electrons dance on your spinal cord. Navajo tribal lands are across the Canyon. Vast Bureau of Land Management tracts are over both the eastern and western horizons, and the trail itself runs through the Kaibab National Forest as well as the park. Wolves once patrolled them all, without regard to human jurisdiction, and helping, with other predators, to play an essential role in the health of the landscape. That became all too apparent once they were no longer on the scene.

No howls tonight. A federal campaign on this plateau from 1906 to 1930 exterminated 816 mountain lions, 7,388 coyotes, 863 bobcats, and the last of the remaining wolves. The absence of predators led to a convulsive jump in the mule deer population, of the kind that biologists call an "irruption." In a scenario that has unfolded in many places when predators are killed off—the Rocky Mountain and Shenandoah National Parks are two current examples—the out-of-control deer numbers may have increased as much as twenty-fold.

The starving animals overbrowsed grasses, shrubs, and young trees. The mauled forest became an early-day case study of environmental mismanagement. A whole generation of aspens, to cite one example, was wiped out during the couple of decades that the deer population rose and crashed. That left, here where I'm hiking, a degraded forest for the networks of insects, birds, and plant life that depend on a full array of tree species of mixed ages.

Wolves were reintroduced at Yellowstone two decades ago, and now there is abundant evidence that the park's landscape, though not fully recovered, is healthier, including its deer and elk herds. Beaver have come back, in numbers. Willow groves, an important wildlife habitat, have returned to what had been eroding, barren stream banks, because they are no longer overbrowsed by elk in the boom-and-bust cycle seen after the wolves were exterminated. Biologists are astonished at the wide range of species that have benefited from the presence of the wolves: eagles, ravens, grizzly and black bears, magpies, beetles, wolverines, and lynxes among them.

So wolves, and the suite of predators they are part of, are essential to complete the web of life. But the beautiful, somnolent cows that have ambled across the National Forest highway on my drives near this trailhead offer something of a contrast. Nearly ten thousand cattle graze the Kaibab National Forest, about 1,500 of them here on the North Kaibab district. They are part of an American herd now estimated at eighty-eight million. The number of rare and endangered wolves on this huge expanse of publicly owned, very suitable wolf habitat in the Grand Canyon region—its national parks, monuments, forests, and sagebrush horizons—is zero.

One female gray wolf did arrive on the North Rim of the park after a 750-mile trek from Wyoming's Northern Rockies in late 2014, to great excitement. She was named "Echo" by her fans. Despite her legal status as an endangered species, she was shot dead by a hunter in Utah soon after. He had thought she was a coyote.

Canis lupus, the wolf, ranged through two-thirds of what is now the United States as late as the close of the 1800s. Among the first waves of wolf migrants from Europe about thirty or forty thousand years before that was a distinct type we now call Mexican wolves. Adapted to a drier, hotter climate, Mexican wolves are smaller than their northern cousins, with a desert-colored calico coat of muted beiges, grays, and rusts. They had dwindled to a tiny number in northern Mexico and Arizona by the 1970s, when the U.S. Fish and Wildlife Service, official protector of our endangered species, began a captive breeding program with only seven animals.

Nearly a half century on, these wolves' survival still teeters. The agency dawdled spectacularly for twenty years before releasing any into the wild. A court ruling finally forced the release of eleven animals into a smallish reserve in southern Arizona. The record shows that lawsuits have continued to provide much of the motivation to push the wolf program forward ever since. After some forty years there are still only ninety-seven Mexican wolves in the wild in the latest count, thirteen fewer than the year before. The project has cost more than $30 million by the most recent estimate. The problem is not that wolves aren't inclined, somehow, to breed and survive.

When I've asked Park Service personnel why wolves have not been reintroduced to Grand Canyon, the answer is that the Fish and Wildlife Service is the lead agency for that kind of thing and that it wouldn't be good for Grand Canyon to take that initiative because "our neighbors don't want wolves here." A few, it is true, do not.

But a 2005 survey by the Social Research Laboratory at Northern Arizona University found other neighbors, about seven hundred randomly selected, representative Arizonans. They were asked whether they support or oppose allowing the wolf "to naturally migrate to its once native Northern Arizona forests and mountain habitats." A full 81 percent supported the idea, and 11 percent opposed it.

A more recent survey, this time a representative sample of Arizona voters, found that they support wolf reintroduction by nearly a six-to-one margin; 67 percent want to give wolves more protection to ensure that their population rebounds; 76 percent agree that the wolf is a benefit to the West and helps maintain the balance of nature; and 62 percent support wolf migration to suitable habitats outside of the zone where wolves are currently allowed.

The Fish and Wildlife Service's regional director for the Southwest is Benjamin Tuggle. He was a wildlife disease researcher when he joined the agency in 1979 and later, as he puts it, "stumbled, and got into management." Part of his job description is to uphold the legal mandate to pull endangered species populations toward recovery and away from extinction.

There is wide latitude in how that is carried out, however, and when. Tuggle views the work as a kind of long-game diplomacy. "One of the things we have to be aware of is the public tolerance of a predatory species," he told me. "You have to do the due diligence, to strike that balance," between the survival needs of the animal and its acceptance by humans who share the habitat. "And a lot of the conservation groups don't want me to say that. They don't want to have that conversation.

"But they have to understand that ... if the public does not accept it, we will never recover these species.... The key to conservation is getting people to believe that they are making a difference and that they are part of the solution." Essential to the process of reintroduction for Mexican wolves is to keep a promise made to ranchers, to remove or kill those that misbehave by preying on cattle, he explained—to accede to the needs of the humans who lease federal land. He refers to that as a "working landscape."

The Fish and Wildlife Service has its own scientists but hires outside consultants to advise its programs too. Wildlife ecologist John Vucetich, a wolf expert, has been an adviser for more than a decade. When a species is listed as endangered, the federal government is required to put together a plan for the restoration of a healthy, self-sustaining population and eventual removal, if things go well, from the endangered list. Vucetich has served on the Mexican wolf planning groups, studies wolf population dynamics around the world, and knows the Endangered Species Act well.

"I think the Fish and Wildlife Service is kneecapping its own efforts," he told me. "And the question is, why? They have the mandate and the technical ability to rescue these wolves. But they take actions that aren't moving in that direction." The delaying tactics, he said, allow the agency to slide away "from having to do the difficult work that might be associated with political conflict. Those delays have lasted many years."

The latest in a long series of lawsuits has forced the agency, after more than four decades, to produce a recovery plan that is supposed to be completed before the end of 2017. The plan will need to be enforced and defended from political interference by the Fish and Wildlife Service. "They sometimes rise to the occasion, and often they don't," Vucetich said. The advent of the Trump era may well diminish those prospects.

"Wherever there are wolves, there are people who hate them," Vucetich added, "but, boy, I've worked on wolf issues all over the world, and they're especially inflamed in the desert Southwest, including in the state government of Arizona." The governors of Colorado, Utah, Arizona, and New Mexico have issued a joint letter to the director of the U.S. Fish and Wildlife Service opposing expansion of Mexican wolf reintroductions within their states. Utah's legislature has provided several hundred thousand dollars to a group that lobbies against any Mexican wolf reintroduction there.

An instructive list of objections to that program has also been drawn up by the commissioners of Washington County, Utah, whose seat is the city of Saint George, not far north of Grand Canyon. Resolution R-2012–1610 states that wolves would prey on cattle, diminish local deer and elk herds, and compete with hunters. It warns too that tourism would suffer. Wolves would "pose a perceived and potentially a real threat" to hikers and campers, so they also threaten tourist-related business interests, the resolution states. "Our tax dollars should not be spent on such programs," it adds, then warns of legal action, though no plans to reintroduce wolves to Utah have been laid.

Some of these arguments are simply misinformed; others are incomplete and misleading. Wolves do prey on cattle. They are a minor factor in cattle mortality, though. Over the most recently surveyed fourteen years of the wolf program, each wolf has been responsible, on average, for the loss of one-fourth of one cow per year. Some individual ranchers sustain higher losses, but ranchers are compensated by the government and by wildlife groups when wolves prey on their cattle.

Contrary to the county resolution, the science research does not demonstrate that reintroduced wolves diminish the populations of game species, and in many cases it contradicts that notion. Reintroduction at Yellowstone twenty years ago has, instead, restored an ecosystem ravaged by an overabundance of elk. Research in Zion National Park, too, has shown that remote backcountry canyons that still have a strong complement of large predators support a richer, more diverse web of small mammals, butterflies, amphibians, and flowering plants. The main stem of Zion Canyon, where predators have not yet returned, is comparatively pauperized in plant and animal diversity.

The number of people fishing and hunting in Utah actually declined by 20 percent from 1996 to 2011. As a house editorial in the *Salt Lake Tribune* put it, "What is Utah's responsibility to the Mexican gray wolf? The answer should reflect the will of all Utahns, not just the ones with guns."

There is no evidence that wolves are much of a threat to hikers and campers except, perhaps, for those who try to feed them. Wolf attacks are extremely rare, Hollywood to the contrary. In places where wild animals are indeed a hazard—black bears in many national parks and grizzly bears at Glacier, the Tetons, and Yellowstone National Parks, for example—tourism thrives, to say the least. The reintroduction of wolves in Yellowstone has generated appreciative excitement—not fear—among millions of visitors since 1995. No attacks recorded.

Those are not the only salient concerns Washington County makes known, however. "It's more our land than yours, because we live here," is more or less the foundation of all these objections. That argument is not without merit. But the taxpayers confidently invoked in the resolution are not the taxpayers who own federal lands and have paid to protect and maintain them for generations. Instead, the resolution really refers only to local interests and their congressional allies. They have effectively paralyzed federal initiatives on behalf of wildlife and natural landscapes, in too many cases to count. The Mexican wolf, which will require a lot more habitat to ensure its survival, may turn out to be one of those.

Specious political arguments aside, wolves do alter the lives of the ranchers they live among in important ways. We're likely to continue to allow grazing on public land (and we'll get around to discussing the wisdom of that pretty soon). So it would be a mistake for proponents of pulling wolves back from extinction not to come to terms with the blood, sweat, and money that we require ranchers to shed if these animals are in the neighborhood.

Carey Dobson, untypical in some ways, runs 1,200 cattle on 310 square miles of leased public land, most of it on the Apache-Sitgreaves National Forest east of Show Low, Arizona. That's an area as large as Shenandoah National Park. He's a hard-working businessman who's moving hay when I arrive—affable and adaptable, but vexed. He's concerned enough about the condition of the grazing lands he leases to periodically commission an outside ecological audit, independent of the evaluations performed by the Forest Service.

Dobson thinks reintroducing wolves is a pretty deranged idea. "I don't want them here," he told me. "They were eradicated a long time ago for a reason. They knew there was no way to stop wolves from preying on cattle and sheep."

But, he added, "I'm not like some other ranchers. I don't look at wolves as the issue. They're doing what they do. They're killers, and they are very, very good at it." The issue for him, instead, is the people who sent the wolves—the federal government, in his calculus—and the way it manages, and fudges, its responsibilities. But if having wolves around is unavoidable, Dobson wants it to succeed, as a kind of non-peaceful coexistence. In this, he is an unwilling exemplar of reasonable accommodation.

His first brush with wolf reintroduction was inauspicious. Something found its way into his horse barn one night in 2002 and chewed a leg off a prized new colt. Dobson inquired repeatedly whether wolves had arrived in his area from where they had been released in New Mexico. He was assured by wildlife officials that they were nowhere around. So with deep regret he put down a guard dog, a beloved Great Pyrenees, the only other plausible culprit. Soon after, someone spotted a wolf on his ranch.

Dobson's family has ranched here for more than a hundred years—four generations, soon to be five. "It's not the wolves' fault," he said. "The government put them on us. They just threw them out there. There were no rules, no communications, no anything." In the years since, Dobson has flown to Washington several times to try to enlist more support. He has worked with state and federal wildlife agencies and the Defenders of Wildlife, an environmental group, to try new management tactics to scare off the wolves and protect his livestock.

Dobson is paid when he can prove that wolves have preyed on a cow, but it can take nine or ten months for the money to come. And the losses go much further. The presence of wolves stresses cattle so they will not readily breed, and the missing calves are needed to regenerate his herd. He has to move his livestock out of harm's way a lot, and he has put up dozens of miles of electric fencing. That and the frequent patrols to fend off the predators cost tens of thousands. Those expenses are mostly uncompensated, in his view.

Elk and antelope were nearby every night when I visited. And a wolf. Four nights earlier calves had been bunched up for branding, and the wolf "killed a calf. Just ate the stomach right out of him," Dobson said. They tried scaring the wolf off with rubber bullets, but "every morning he's out here chasing things around. My dogs go crazy, but we keep them in. We've got used to living with wolves." By the fall of that year, he would lose two cows and six calves to wolves. These were officially confirmed as wolf kills, he told me. Only five cattle died of other causes over the same period.

Dobson is allowed to kill a wolf on his private property if he's in danger or if it is attacking cattle there, but he said he wouldn't chance it. "I'd never shoot a wolf. I love my ranch. I love working here. If I couldn't prove that it was justified, I would be finished here. I'm not going to jeopardize all this. I don't think a rancher is dumb enough to try something like that."

But the Endangered Species Act is not the misbegotten child of the federal government, dreamed up by bureaucrats. That's just easy rancher rhetoric, eagerly stoked by corporate and ideological allies who stand to gain from a weakened federal presence. The act is the codified will of the citizens of the United States, including heavy majorities of the citizens of Arizona, that wolves should roam here on public land. There's a legitimate question just the same, though: whether, once we've visited these problems on Dobson and, potentially, hundreds of other folks, we're really supporting endangered species programs adequately and consistently.

One way to approach that question is to ask, if the losses stay more or less where they are, can Dobson stay in business? "Oh yeah. I do it. I am a public-lands rancher. I love it," he told me. "I'm used to them. I know how to deal with them and live with them. I have to. But I'm dishing out so much money, it's not fair."

Part of the equation here, unannounced but clearly understood by Dobson, is that he is allowed to lease enormous tracts of public land cheaply—at a rate subsidized by the federal government—and profitably. Looked at squarely, that is his inducement to put up with the considerable burdens of regulation and of wolves. "Ranching with wolves, I mean, it's a tough job," he said. "But it's what I have to do."

It would be too easy, as the wildlife ecologist John Vucetich notes, just to spotlight the compromised efforts that Forest Service, Bureau of Land

Management, or Fish and Wildlife Service officials often make to protect endangered species on public lands. They try to placate one side of this conflict, then the other, seemingly without clear priorities other than bending with the political breezes. That's because those officials rely ultimately on public support to carry out the laws. For now, cheerleading for the Endangered Species Act, for example, is not heard at the same volume as the hostility that also registers in campaign donations and votes.

One tactic that forwards the work of dismantling the Endangered Species Act is to conjure up a powerfully misleading focus on the single species whose status is in question. Then the one animal is weighed, on the golden scales of what passes for common sense, against economic interests that will allegedly suffer if it is protected—they are usually reduced to "jobs." This trivializes matters effectively. More so when the species is, unlike the wolf, not especially charismatic.

The extinction issue is not just spotted owls, snail darters, or checkerspot butterflies. It's that these are alarm bells, none too faint these days, signaling that the natural systems our own species depends on are unraveling. I once heard a rueful Fish and Wildlife Service official recount his appearance at a public hearing in Southern California. He was there to explain a program to protect the Delhi Sands Flower-Loving Fly. Real estate developers packed the audience. They all came with flyswatters. Hilarious. Later, a local congressman submitted a bill to dispense with the Endangered Species Act and the insect too. Local headline: "Baca Bill Aims to Swat Bothersome Fly." Right.

But the larger benefit of species protection is that entire habitats are also preserved from destruction, along with their tapestries of natural systems and other species, including, in this instance, a pocket mouse, a burrowing owl, and the Mormon metalmark butterfly. "It's the ecosystem, stupid!" That would make a fine campaign slogan. As many as ten other species are unique to the Delhi Sands terrain, but 97 percent of it no longer exists.

In fact, the Endangered Species Act itself is endangered. One conservation group that troubles to keep track noted eighty-one legislative attacks on the act in Congress during a single recent year, and the rate is accelerating. Interest groups such as the American Petroleum Institute, the Western Energy Alliance, the National Rifle Association, and the U.S. Chamber of Commerce promote such initiatives, as well as their sponsoring politicians. One soldier in the anti-ESA ranks is former Congressman Ryan Zinke of Montana. On twenty-one occasions in which protection of species was at issue during his two years in office, he voted against it. He was named as secretary of the interior, the new boss of all national parks and public lands, and the ESA, as the Trump administration took office.

"It's not entirely clear to me how many citizens really appreciate that law," Vucetich told me. "As a society, maybe we haven't quite figured out what kind of relationship we want to have with nature, and whether we want to have wolves," though the health and survival of wildlife has solid backing whenever Americans respond to polls on such questions. For Mexican wolves, the water treading can't go on indefinitely, however. "There is a ticking clock," Vucetich said. "If the population is kept too small for too long, you can accumulate enough genetic damage that you can't recover from it. That's absolutely certain."

That question engages the expertise of Richard Fredrickson, a population ecologist and geneticist who is also a member of the wolf recovery team. "Not many of those wolves die of old age or natural causes," he said. "Most die of human causes or are removed for perceived indiscretions such as preying on cattle or approaching houses too closely." The Mexican wolf program thus far, he added, has until recently been a disaster in that regard. For awhile, the Fish and Wildlife Service yielded control to the Arizona Game and Fish Department. So many wolves were removed from the wild that "they seemed intent on running the population into the ground, and they did a pretty dang good job of it," Fredrickson said.

That policy has been reversed, for the time being, but the wolves are far from safety and stability, Vucetich and Fredrickson agree. Illegal shooting or trapping accounts for more than half of the recorded deaths of Mexican wolves.

These dim prospects have everything to do with the regional politics and the culture of vacillation in the national leadership of the Fish and Wildlife Service. Two thousand miles away a kindred program to reintroduce a different animal, the endangered red wolf, is also in disarray. Its wild population is in free fall after decades of careful nurturing, mostly because of gunshot fatalities. Also, the coastal North Carolina wildlife refuge the red wolves inhabit is disappearing because of sea-level rise caused by climate change. So the recovery of both red wolves and Mexican wolves remains in peril, biologically and politically.

Let's change direction, though, and pretend that I've been busy relaying the thoughts of some outliers—wildlife biologists cherry-picked for their disgruntlement. The Fish and Wildlife Service regional director Benjamin Tuggle observes, after all, that wildlife biologists out in the field do their research and reach their conclusions, but they are not responsible for the tough decisions.

The overriding question, however, is how those decisions are made, and on whose behalf. The Union of Concerned Scientists recently surveyed several thousand scientists at four federal agencies to ask various questions about the scientific integrity of their government employers: the Fish and Wildlife Service, the Centers for Disease Control, the Food and Drug Administration, and the National Oceanic and Atmospheric Administration. Fish and Wildlife Service scientists reported by far the most frequent outside political interference in their work. Three out of four of those scientists said the agency pays a high level of "inappropriate consideration" to political interests. Nearly half said it kowtows to business interests to a high degree. "Managers should actively solicit input from field biologists and not cultivate a 'culture of fear,'" one of them wrote.

I asked for a response from the Washington headquarters of the service, custodian of the fates of endangered species and our national network of wildlife refuges, expecting a spirited rebuttal. "The Service is fully committed to the highest standards of scientific integrity," was the official response. "We will carefully review the information in the survey and continue our commitment to ensure broad awareness, understanding, and implementation of the Department of the Interior's Science Integrity Policy."

The new plan for Mexican wolf recovery, delayed for decades, could finally map a secure path away from extinction. If so, its success "will depend on Fish and Wildlife Service folks in DC," Fredrickson told me. Tuggle, the agency's regional administrator, denied that politics affects regional decisions about endangered species. He added, though, that policy decisions usually come from Washington.

The jaguar's historical range spanned at least the distance from California to Louisiana and perhaps farther east. It is magnificent, the largest species of cat in the Western Hemisphere, and can be found as far south as northern Argentina. A small and diminishing population of jaguars persists, barely, in the mountainous region of the state of Sonora, Mexico, despite centuries of human encroachment and hunting. That is the likely source of the four jaguars that have entered Arizona since the 1990s, despite the walls, fences, and other measures we have taken to thwart human migrations.

For its part the federal government has pursued its role as guardian of these wild wards with marked reluctance. At first the Fish and Wildlife Service declined to consider jaguars to be an endangered species—or even a resident native species—at all. Cornered by a lawsuit, the agency acknowledged in 1997 that at least sixty-four jaguars had been killed in Arizona since 1900, and that a few still survived, "at least as an occasional wanderer from Mexico."

The species was finally designated as endangered, but then the Fish and Wildlife Service refused to map out a protected "critical habitat" for jaguars—one of the requirements of the Endangered Species Act. Ten years and a couple of additional lawsuits later, the agency continued to resist developing either a recovery plan or designating critical habitat.

To critics the agency's subsequent maneuvers seemed to rival the jaguar itself for crafty evasion. Federal district court judge John Roll, ruling on the next lawsuit in 2009, tossed the rationalizations aside. After a comprehensive review his order stated that the Fish and Wildlife Service's continuing refusal to protect jaguars was not based on the best available scientific evidence and was not consistent with the Endangered Species Act, the agency's own regulations, or previous court rulings in similar cases.

Roll told the Fish and Wildlife Service to reconsider its decisions not to prepare a recovery plan and at long last to go ahead and designate critical habitat. Four years later the recovery plan was still not completed, but at least the habitat was marked out: 1,200 square miles of rugged southern Arizona desert and mountains.

Then a U.S. Fish and Wildlife Service administrator decided—overruling the findings of his agency's own biologists—that it would be fine to approve the development of an impressive new copper mine, owned by a Canadian company, right there in the new jaguar "critical habitat." Two-thirds of the mine's total of eight square miles were to be on Coronado National Forest land. Its terraced open pit would be more than a mile in diameter and three thousand feet deep. The mine would operate for twenty to fifty years in an area where one of the endangered jaguars has recently been tracked. This is habitat for a range of other endangered plants and animals too. The mine, the jaguar, and the other species are, as of this writing, in limbo, while the mine owner awaits improvement in the international minerals markets, and other federal agencies weigh their options under the Donald Trump administration.

The Arizona Game and Fish Department has also had a hand in the prospects for jaguar survival. A male nicknamed "Macho B" was snared "accidentally" in 2009, according to the department, then fitted with a radio collar. The animal fell ill soon after, traumatized by the capture, which had been carried out using questionable methods. "He died a very slow and painful death for 12 days," one employee wrote later. "His stomach was empty, he was dehydrated. I can't imagine what he went through." The jaguar was then caught and euthanized.

But an *Arizona Republic* newspaper investigation found that the initial capture was carefully planned rather than accidental, that the department had lied strenuously in a subsequent cover-up, and that at least one federal jaguar expert was also implicated.

Tony Povilitis is an Arizona-based wildlife biologist who works on endangered species projects around the world. He has also maintained an interest in the future of the jaguar, particularly in the United States. "I did work in Kenya," he told me. "The Kenyan Wildlife Service can be outspoken and has gotten in some tough fights. They've stood up for wildlife, even going against powerful special interests. In this country, we need much more of that.

"As a democracy we should be robust enough to have those agencies that are supposed to stand up for wildlife actually do it. The bottom line is that they're there to serve the American public and to protect America's wildlife, not just hunters and ranchers. It's pathetic that our national agency that is responsible for recovering endangered species can't make the decisions that are needed."

In Arizona, he said, those same constituencies also prevail, and "the rest of public is pretty much shut out." It is true that the Arizona Game and Fish Commission, for example, includes no biologists, ecologists, endangered species specialists, or environmentalists. As of 2016, four out of five commissioners were sport hunters; the fifth was a retired law officer of unknown affinities who did not respond to inquiries. Three are publicly avowed members of the National Rifle Association. Another, a member of the Yuma Rod and Gun Club, dropped his membership in the NRA several years ago.

Animals like jaguars need to be protected from being shot or harassed or having their habitat paved over, Povilitis told me, but they also need protected corridors to migrate, in light of global warming. He said he has escorted U.S. Fish and Wildlife Service officials to corridor areas in Arizona well suited to connect patches of protected wilderness. Their response was that habitat corridors are not a priority.

The agency "moves at glacial speed," he said. "Ultimately, if this is ignored we will start looking more and more like Southern California, where all the corridors are closing or already closed. We're talking about an endangered species that the law requires us to recover, but you can't have a recovery unless you have protected corridors."

One essential set of hoped-for corridors extends toward the Grand Canyon—higher country that affords more sources of water and an abundance of deer and other prey for wolves, jaguars, and many other species. "It's still quite possible that this vision of having an interconnected system of open space and habitat would work, and in the event of serious climate change would certainly help to maintain different species of wildlife," Povilitis said.

And there is no reason, he added, why the missing species should not be reintroduced to large areas of Arizona's national public lands, including those of the Grand Canyon region. "These animals have a fair amount of flexibility in terms of their prey," he said. "There's an excellent chance of success. It's clearly doable if you have the will to do it."

The Endangered Species Act is complex, though its aims, the courts have found, are clear enough: to decide which U.S. species are in urgent need of protection, to grant them protected habitat, and to devise a plan for their recovery that will allow them to survive into the future, unaided by humans. The Supreme Court has stated that "beyond doubt ... Congress intended endangered species to be afforded the highest of priorities," and the "plain intent of Congress in enacting the statute was to halt and reverse the trend."

But the U.S. Fish and Wildlife Service has reinterpreted the act. It has decided that the original ranges of endangered species are no longer to be taken into account when risks to the survival of those species are calculated. The agency now declares itself obligated to restore them only to the "critical habitats," however meager, that it agrees to—or is forced to—create for them.

The wildlife ecologist John Vucetich and a colleague, Michael Paul Nelson, have characterized this as a dangerous retreat: "This means that as long as a small, geographically isolated population remains viable, it won't matter if the animal or plant in question has disappeared across a vast swath of its former habitat. It won't qualify for protection," they have written.

The two authors invite us to imagine the result if this approach had been used when the bald eagle was near extinction five decades ago. Our national symbol might have been listed as endangered only in the few places where it still endured rather than most of the lower forty-eight states, its former habitat. "Today, the return of the bald eagle is one of the great successes of the Endangered Species Act. The bird is flourishing in the very areas where it had been wiped out and is reasserting its position in the ecological order that was disrupted by its absence," Vucetich and Nelson write.

Povilitis told me, "If we're serious about restoring a native species, we define that as establishing an abundance of that animal, to a range that is at least somewhat comparable to what it was historically. As long as there is habitat out there, that should be the goal."

The Fish and Wildlife Service's shriveled aspiration calls for something akin to a set of outdoor zoos, with severely dimmed prospects for enduring, healthy populations of wild species. Certainly that blueprint does not account at all for the survival factors emerging in Arizona and throughout the United States as climate disruption proceeds. Despite that threat, we can only imagine, rather than see or sense, many wild species—the ghosts of animals that once lived on what are, at least for now, federal lands, including Grand Canyon.

CHAPTER FIVE

Tusayans

The national parks are not in the least degree the special property of those who happen to live near them. They are national domain.

> Freeman Tilden, "Father of Interpretation,"
> National Park Service

On a whim I catch a ride out to Hermit's Rest. It's a National Historic Landmark, a shapeless, century-old rubblestone store that nearly disappears into its surroundings. That was the architect Mary Colter's fine plan. Here also is a trailhead, layered in deep pink dust that puffs up miniclouds with each footfall. Hot today, even in spring, even up here high on the Canyon rim. Midmorning shimmers off the rocks and the relics.

The Hermit Trail winds down nine hard miles from here to the Colorado, and the usual sign is posted at the edge of the descent to warn the heedless and the smug: don't try for a long haul unless you're truly ready, and certainly not to the river and back in a day. Take a lot of water. People get in trouble on this trail.

As if to punctuate the message, a quite miserable set of parents with their two early-teen daughters hobble and sweat past, flashing rueful smiles. As for me, I know I couldn't go far on a half-empty water bottle—and then, I'm more or less out of shape too. Mostly more. Pretty day, though, on this route known for its wide views and knee-pounding inclines. Some shade along the rocks and some pleasant stops. Every time I pause to think about turning around, I don't.

A mile and a half down, the trail finally relents to a level stretch. It is past noon. Around the bend ahead a stunning amphitheater climbs to the horizon on my left—bleacher seats for the gods. On the right the Hermit Creek gorge zags down toward the Colorado with a certain tractive grip. The map shows me a destination called Dripping Spring, which does not seem to be much farther on.

This didn't end particularly well. Hours later I was waterless through the last long furnaceous grind back to the top, pacing off, toward the finish, a meager hundred footsore steps between panting rests. At least, I told myself, I did not enlist in the legion of hikers who have to be rescued here each season.

The miscalculation was useful, though. It led to a blunt reminder of what water means for this stark landscape—at the far end of the trail to Dripping Spring. That trickle emerges from some fracture deep within a cul-de-sac of tall, hot sandstone. It waters a mop of maidenhair fern hanging from a ledge overhead. Then its silver thread falls fifteen feet, thumps your grateful skull or fills your bottle, and splashes the floor of a shady alcove where hikers drink and gather their resolve.

Below there it sustains a rich green patch of grasses and shrubs, startling against their arid surroundings. The spring can wane unpredictably, though. Then its strip of green turns sere. My timing was fortunate. Without that trace of water, I might have withered more seriously, too.

Some of these are seasonal water sources, some perennial. Some are fed by rain and melting snow that penetrate down through unknowable mazes of fissures, joints, and aquifers in the strata, perhaps thousands of feet above or tens of miles off. The water you stop to guzzle on a parched hike may have taken quite a long while to find a path to you. Isotopes sampled at Canyon springs show that some of the water has been locked away, moving at an exquisitely languid pace within the rock for as long as twenty thousand years before emerging.

"You don't want to take risks with that water," a park resource manager, Jan Balsom, told me. "The aquifer situation is so fragile. We are looking at external threats from mining, development, and climate change that will affect both seeps and springs. Anything that degrades them can have a devastating effect."

About nine hundred seeps and springs are known within the Canyon. They occupy only a hundredth of 1 percent of the land area, but they account for most of its biology: the margins of these mostly tiny flows give life to as many as five hundred times more different species than the surrounding terrain.

Water defines our own survival prospects too. You can feel that in the dry breezes that kick up swirls of grit and dead leaves at a little old community called Tusayan, a small cluster of homes within the park. I spent an afternoon there with Balsom and park archaeologist Ellen Brennan, who know a lot about the twenty or thirty citizens of this tiny refuge and their history. Between them they have put in more than sixty years at Grand Canyon. They often visit here at the Tusayan Ruin, which has been uninhabited for eight centuries.

Ancestral Puebloans lived in these dwellings of rock, mud, and thatch and also within a complex culture—language, religion, conflict resolution—that had to respond to its physical environment to survive. Some of the social energy, the rules and ritual, must have been devoted to coping with scarcity. There is little surface water up here to speak of, so the inhabitants had ceramic vessels to store rain or melted snow or to carry it from springs.

"There was never enough water to drink, because it was a precious resource and they knew it was finite," Brennan says. "Their social mores would tell them how to relate to that environment and to each other— what is acceptable, what isn't. Things like, drink only enough water to satisfy your need. Don't overuse the environment, so we can continue to live here." The primary force behind the gradual abandonment of the region, as its inhabitants withdrew to the east, was a long, killing drought.

"They were no dummies," Balsom says, looking out over the remains of storerooms, living spaces, and a circular kiva, a kind of church and community center—the rock husks of social history, gradually disassembling. "They survived here for twelve thousand years and their tradition continues among the Hopi and the Zuni. We've managed about two hundred years." And just now things may be pretty dodgy for us in terms of continuity, she observes. Mapping the comings and goings of water under the surface of the land here can help us figure out how to use it sanely. That expertise, hydrology, is a major part of conversations about the future of life in and around the Canyon now. There's global warming and its projected megadroughts to consider, and their impact on those life-giving springs. Even more imminent: another, much younger Tusayan and its quite different culture has to be reckoned with. The way decisions are made about land and resources in that hamlet may well register in the Canyon too. "I guess one of the very first things to say about the new Tusayan is that it's not in tune with its environment at all," Brennan tells me. "It doesn't know its carrying capacity."

Tom DePaolo considers some of that, while presiding over breakfast at a place along Arizona 64. It's the main thoroughfare of the other Tusayan, a gateway town just a mile south of the main entrance to the Grand Canyon—a modest lineup of hotels, pizza and taco parlors, and cowboy-themed steakhouses. Its population is about six hundred—plus the 5.5 million park visitors who come through each year. They are the town's only real resource, its reason for being. They float through on a river of cash, and tiny Tusayan is a port of entry with some big and thirsty plans.

Gruppo Stilo, an Italian developer and the town's biggest landlord, is the author of those plans. DePaolo has been Stilo's agent here for a quarter of a century, as first one, then another, development initiative has run aground. He is said to be a charmer, even by some of his detractors.

He is laid back this morning, bemused, a storyteller. "I've always been in commercial real estate development," he begins. It was hotels, until that industry cratered for awhile with mid-eighties tax reform. "Then I got involved in some factory outlet stores, which is actually what got me up to the Canyon.... I'm sure you've heard about all that we were going to do to destroy things up here," he jokes.

He confesses ruefully that way back then he knew no better than to think he might buy fifty acres and open a Ralph Lauren outlet in the national park. If a pigeonhole seems to form around this resume, though, it's a little hard to fit the man himself into it at times. He knows the park scene better now, after dozens of trips to the bottom of the Canyon, on the trails, and on the river over the years.

DePaolo makes the case that his employers are a different breed, with the long perspective of ancient Rome in their developer DNA. Among several locals who made junkets to Italy at Stilo's expense, he recalled, was a Hopi tribal leader. "He'd lecture me about how they've lived here for a thousand years. I took him over to Florence. I told him, 'After a thousand years, you don't get even a plaque. You get a tube of Clearasil and a note: Come and see me when your teenage acne has healed'"

DePaolo added that his employers consult those same broad horizons. "These guys understand something the U.S. developer certainly doesn't, and to some extent, I am embarrassed to say as a citizen, we don't appreciate special places, cultural treasures.... They know that there's one Grand Canyon in the whole world. From a pure financial standpoint, if something happens that they're able to be part of it, it's forever—especially if you do it right."

Getting a shot at forever has meant some hardball in the here and now, though. Stilo tried with a large-scale, mixed-use development plan in 2000 that won support from some environmental groups, Native American tribes, the Forest Service, and the Park Service. Opponents charged that, despite the developer's promise to pipe in water over long distances, the plan would ultimately threaten Canyon springs. It was defeated in a county-wide zoning referendum.

At the time Arizona law specified that only communities with a population of 1,500 or more could incorporate. It's a reasonable lower limit.

Communities smaller than that, the logic goes, cannot support the administrative functions necessary to properly govern a town. So they are governed by counties instead. In 2003, however, the state legislature amended the law to read that "a community within 10 miles of the boundary of a national park or monument that contains a population of 500 or more persons" could also incorporate. Tusayan was the rather obvious target of that change. Only Tusayan business owners or their representatives appeared in support of the bill at a legislative hearing.

Soon after, a local group convened to study incorporation, then disbanded without a recommendation. Five of its seven members said privately, though, that the town is too small to shoulder its own government and probably could not do it in an ethical way, either. The owners of only a few businesses provide all the jobs. Locals elected to a town council might be burdened with conflicts of interest over a range of important decisions. They would either have a financial stake themselves or be employed by, and thus beholden to, the people who do.

There were disputes about whether Tusayan even had the requisite five hundred souls to qualify to vote on incorporation, but a campaign was mounted, with financial support from Gruppo Stilo. It afforded free trips to Italy for some community leaders to see the firm's operations there. The vote was narrowly against incorporation anyway.

Very soon after, another vote was scheduled. This time Stilo recruited a well-connected Scottsdale public-relations firm, the Policy Development Group. Its representative, Andy Jacobs, is also at our table this morning. By way of introduction Jacobs reminds me, as his eggs arrive, that his firm "ran the incorporation campaign, which Stilo funded."

It's not breaking news, this blithe summary, but it distills nicely. Stilo finds a work-around: the county voted down the campaign's plans, so it fronts some cash and creates a town instead. "What got us excited about incorporation was that all of a sudden you've reduced the world that controls your future to where you can sit down, one on one, and explain," DePaolo tells me. "That's feasible." And this time the incorporators won. Some locals received cash "win bonuses" from Stilo for working on the campaign. A couple of them were later elected to the new town council.

One of Stilo's business partners, another major local property owner and employer, is Elling Halvorson. He owns several businesses in tiny Tusayan, including Grand Canyon's largest helicopter air-tour service, based here and in Las Vegas. The ultimate success of Stilo's incorporation campaign can be measured in the affinities—for both Stilo and Halvorson—of the town council members in office during my visit.

Tusayan's mayor, who is also a council member, manages a local hotel owned by Halvorson. Three other council members work for the Halvorson air-tour operations. The fifth council member manages a local trailer park, jointly owned by Halvorson and the Stilo Group. The town has only three full-time employees, including an interim town manager who works out of an office at the airport, well within earshot of the tourist helicopter and airplane fleet.

Tusayan's town council members have not violated state conflict-of-interest laws when voting on Stilo-related matters, according to Town Attorney William Sims. But even if everyone swears that they aren't influenced by their connections with these two major landowners and developers, the situation is bad for public confidence in the town's government and its probity, attorney Robert Wechsler told me. He is director of research for the nonprofit organization City Ethics and has focused on this subject nationally for a decade. "Appearance is everything—it's all we citizens have to go by," he said. Tusayan presents a case in which the benefit that a town council member receives from decisions that favor his employer is indirect. But it is a benefit nonetheless. "If you vote for your employer's economic interests it may well help your job prospects, and if you vote against them, it could easily hurt. Anybody in that position may be conflicted," he said.

Once incorporation occurred, "then it's a town being managed for the long-term economic benefit of the owners versus what is in the best public interest," former Grand Canyon superintendent David

Uberuaga remarked. "Rationality and good decision-making just go out the door."

Why dwell on the involuted politics of this town of a few hundred folks? Because Gruppo Stilo has proposed a venture that may be underway—or completed or shelved—by the time you read this. It includes 2,200 homes, hotels, a dude ranch, a Native American craft outlet and cultural center, and a spa. It all weighs in at three million square feet of commercial space. If this were a shopping mall, it would be the third largest in America.

A Park Service analysis found that water consumption at something like this version of Tusayan would quadruple. In spirit, the proposal gazes steadfastly at Phoenix or Las Vegas and turns its back on Grand Canyon—except for the park's tourist traffic and revenue stream. Expansion of the local airport to accommodate major airlines is also in prospect. Because of the helicopter tours, it is already the third busiest airport in Arizona.

More visitors, staying longer: that kind of growth in Tusayan means "the pressure of more cars, more people, more water use, more housing, more infrastructure, more wastewater. We can't sustain it. It diminishes the resources the park was set aside for," Brennan, the archaeologist, told me. The park has severe trouble managing its millions of visitors now, including its maintenance backlog of $371.6 million.

Grand Canyon already has the highest hotel occupancy rate in the country, according to Uberuaga. Often every hotel room, parking space, and shuttle bus seat is full: "When do you say enough is enough— the visitor experience is being impaired and we need to start scheduling the visitation?"

Concerns about increased overcrowding from the big Tusayan proposal are so intense that they can obscure the potential impact of noise pollution, light pollution, and visual decay, not much more to be savored than the appearance of a pot-banging, neon-clad burger mill at the nave of a cathedral. "Do we want this to be a place where people have three million square feet of high-end shopping?" Brennan asks. "Is that the experience that we want them to have before they come and look at a World Heritage Site that is supposed to be unchanged for all time?"

Water rings the loudest alarm, however. Pre-expansion Tusayan depends on wells that already pump enough water to concern park scientists about their impact within the Canyon. "Every decision we have to make about development is based on the question of what the water future looks like," Uberuaga said. "What is the impact of external development pulling water resources out?" The obvious research project would be a multimillion-dollar network of test wells to map and monitor the sources of the Canyon's water supply. All of that just might tell us what kind of impact Tusayan's pumps and spigots already have at, say, a place like Dripping Spring, or how things might go when climate change bears down on the subterranean water supply. "We don't have the financial resources to study that," Uberuaga told me. "We are squeezed continuously." No other interested parties have volunteered the money, either.

I asked DePaolo what his employer intends to do about water for the new development, but no decision had been made. There is talk of bringing water in by train or recharging aquifers with recycled water. Drill? "Legally, we have the right," he affirms. "We've not hidden anything," he added. "I've been very candid with people. But what I don't want to do is say something we don't know."

DePaolo and Jacobs are here for a meeting later this morning with the town council. Two Stilo agents from Italy have flown in for the occasion. They will all discuss the town's application to the Forest Service to allow a road easement on national forest land. The roads are needed to access Stilo properties that will be part of the new development. I invite myself along to the meeting, but it is closed to the public.

Tusayan's town council voted to apply for that easement and forwarded the development, as it happened, despite having no idea from Gruppo Stilo where the water supply would come from. It seems clear that in the absence of a real town government here, with its own staff of independent public servants, Stilo has been willing to fill the vacuum. Just as clear: key resources at Grand Canyon National Park are, to an astonishing degree, in the hands of the five town council members—let's assume their good intentions—and, of course, those of their corporate mentors.

Tusayan is part of a pattern of swelling commercial enterprise along the borders of many national parks, forests, wildernesses, and monuments that could once rely on wide buffers of undeveloped rural land. Those now erode, sometimes very quickly.

Another commercial thrust has been proposed on 420 acres outside the park, at the confluence of the Colorado and the Little Colorado Rivers. Called Grand Canyon Escalade, it features a gondola tramway from the Canyon rim to the river; the 1,900-foot Riverwalk along the floor of the Canyon; the Confluence restaurant; "themed cultural and historical recreation entertainment;" and shopping experiences.

Grand Canyon is the second-most visited of the national parks. The third-ranked is the oldest park in the system, Yellowstone. One of its gateways, the town of Jackson Hole, is now burdened with congested roads, car exhaust, and condos. "The dirty little secret" of Yellowstone and nearby Grand Teton National Parks is that they have limited occupancy, Jonathan Schechter, executive director of an economic think tank, told a reporter. "You just can't cram too many more people into Yellowstone. If a lot more people do come, where are they going to stay? They are going to have to spill over into gateway communities. It's not an issue being widely addressed."

The town of Estes Park, Colorado, which has accumulated along the margins of Rocky Mountain National Park, shows some of the worst traffic congestion of any gateway. High-season gridlock is so ferocious that it can take twenty minutes to travel five blocks. A similar car cloud has enveloped Great Smoky Mountains National Park. It is the busiest in the system, with twice as many visitors as Grand Canyon. Its gateway town, Gatlinburg, Tennessee, is a renowned eyesore, a wild aggregation of waffle palaces; souvenir outlets and gas stations; a German-themed skyride, amusement park, and ski resort; an aging space needle; and the Dollywood theme park nearby—a near-absolute disconnection from the natural heritage of the park that brought the town into being. Gatlinburg's motif, too, turns resolutely away from the magnificent Smokies and toward the urban free-for-all that visitors used to come there to escape.

Occasional acrimony from local stakeholders is even older than the park system itself. It dogged the creation of national parks at Grand Canyon and at Yosemite. Grand Teton National Park took decades to establish because of local opposition. Even an outright donation of thirty-five thousand acres to help push the park along was originally opposed by sheep ranchers, Forest Service personnel, business interests, dude ranchers, and hunters. The visiting superintendent of neighboring Yellowstone National Park was said to have been nearly run out of Jackson Hole, the town near what is now the Teton park. At some point, however, even more alarmed by the proliferation of shabby commercialization and some proposals to build dams, most of its former opponents aligned in support of the new national park.

Some community leaders around Shenandoah National Park in Virginia also have a long history of animosity, partly because sections of the park were created through eminent domain proceedings. In the past, local governments or business groups have opposed antipollution measures, nonprofit conservation funds to purchase additional park land, and even the park's efforts to cope with overcrowded parking at trailheads.

It may seem quite natural when some (by no means all) business owners and developers in gateway towns like Tusayan identify their own plans and interests as none of the park's business, even while they tend to see the parks as existing for their benefit. The late author Freeman Tilden, sometimes called the "father of interpretation" for the national parks, addressed the error in an essay written in the mid-1900s: "The national parks are not in the least degree the special property of those who happen to live near them. They are national domain.

Yellowstone and Yosemite belong as much to the citizens of Maine as to those of Wyoming and California; Isle Royale to the New Mexican as much as to the people of Michigan."

Luther Propst is a veteran of the development-versus-conservation wars. A lawyer with a planning degree, he founded the Sonoran Institute, which brokered and mediated real estate development projects in environmentally sensitive areas for twenty years. He is a coauthor of the book *Balancing Nature and Commerce in Gateway Communities.*

"Some of the more hard-line groups would think I'm too much of a moderate," he told me. "I've always been seen as a compromiser and a person who is working more on community issues instead of pure biodiversity issues. One guy said my institute was the ladies' auxiliary of the conservation movement." Naturally, he rejects the characterization. I asked him what he would have proposed for Tusayan if it were a blank slate instead of a war zone. What should be there?

"What I would like to see is that as many people as possible get out of their car in Williams and take the train to the Canyon and leave the car at the interstate or in Flagstaff or Tusayan and take the bus in. National parks in this century have the unique opportunity, really the responsibility, to help people get out of their automobile, so they are more likely to have the kind of cathartic, life-changing, 'Aha! Nature matters!' experience that doesn't happen behind the windshield. If I were the king, Tusayan would look quite a lot different than it does.... In my world, you wouldn't drive your car right up to the rim."

He'd also move housing that now exists inside the park out to Tusayan and make that town more of a community, where children could bike and walk to school. He would do much more to make Tusayan's major employer-landowners provide decent, affordable housing for employees. Three million square feet of commercial space wouldn't be on the horizon. "Do that shopping in Williams or Flagstaff," he said.

I asked Propst if Arizona could impose those controls on private developers. "Oh, absolutely. Nothing required the legislature to allow

Tusayan to incorporate," he replied. "County plans can limit development to very low density in isolated areas like that one. In almost every development of this kind, that happens; there are significant public subsidies: tax preferences, for example.

"The issue with planning is political will," Propst summarized. "If Coconino County and the Arizona legislature did not want to encourage development in Tusayan, there's all sorts of tools, including land-use planning, discretionary zoning changes, and, importantly, through eliminating or reducing the significant public subsidies that fund isolated development.

"Who pays for highways and the airport expansion? They could stop it if they wanted to. If landowners go to Phoenix and throw a lot of money around that says they want to incorporate a small place like that, they get the changes they want. That's the way the Arizona legislature works. It has absolutely no interest in being a partner with the park."

Propst spends most of his time living and working in Jackson, Wyoming, now, in the shadow of the Tetons and near Yellowstone. He has seen large-scale real estate development in neighboring counties, where there's plenty of investment capital and "a history of anything goes" without regard for environmental impacts.

"I'm more concerned about those communities following the Tusayan model of rapid, large-scale growth and tourism around these national parks that is less related to the resource and more related to amusement. Jackson Hole could become an amusement center, not tied to the park, the way Gatlinburg has.

"Fifty years ago Gatlinburg was very much the place you went to go see the park, and maybe a bear, and look at the night sky. Now the much bigger attractions are the music halls and the putt-putt courses, and I worry about that happening in Jackson and in Tusayan." He calls it the "commotion of come-here." It's the path that really worries him. "It may become like Vegas," he said.

But ruinous development on the edges of national parks is not an inevitability, even if local and state governments doze or collude. In the

1990s the Walt Disney Company put forward plans to spread a theme park called "Disney's America" over thousands of acres, among the Civil War battlefields and horse farms of northern Virginia. It had the support of the state government but aroused opposition, locally and nationally. At one point, three thousand protestors from around the country came to the region to march and chant, and litigation was on the way.

Disney halted the project. "We do feel good about it," James McPherson, a Princeton history professor and a leader of the campaign, said at the time. "Disney recognized what it was costing them in terms of image, public relations, and the potential for a long, drawn-out controversy and lawsuits from environmental groups."

The tools for better planning for gateway communities include patient diplomacy on the part of park officials. These days, though, patience can seem like the only tool they have. Diplomacy works best when there is some muscle and maneuver available to both sides. At Grand Canyon and other parks, those advantages can seem to belong to the developers alone.

In our distracted and pauperized news media, park officials, like Grand Canyon's Uberuaga, are reduced to occasional attempts to win a public hearing about what's going on. Uberuaga caricatures a meeting with the governor of Arizona that gives a sense of the political pecking order: "I walk in to see the governor greeting the mayor of Tusayan. 'Oh great! Great! How you doing? Hug, hug, oh I haven't heard from you for a week!'" Then the governor turned to Uberuaga: "'... and, and who are you?' And I was in my uniform!"

Historically, national parks have sometimes had enough national political support to fight hard to protect their resources. Under those circumstances a balance of power with local interests can be more easily struck. When Grand Canyon was made a national preserve, there was diplomacy. There was also a political brawl, not to be welcomed but not run from either, by the Theodore Roosevelt administration.

When Tusayan's proprietors demonstrate an easygoing mood about the prospect of drilling for scarce water, about overcrowding or inappropriate development, then what if park officials were in a position to return the disfavor? Buried in a set of twenty-year-old park planning documents is one proposal that would get rid of most of the noise and pollution of auto traffic in Grand Canyon. It won enthusiastic support from frontline staff at the time but was rejected by administrators. It would close down the main park entrance that brings traffic through Tusayan and reroute it to a staging area well east of the town and outside the park.

Visitors could ride a shuttle in to sightsee or backpack or find their lodgings, as Luther Propst suggests. In that scenario the town of Tusayan would be a cul-de-sac, cut off from the park. Its proprietors could take their chances with that new role: not as a gateway to a glorious national park but as some species of shopping destination, with condos.

That strategy, and similar leverage for the rest of our national parks, may seem an unlikely outcome. And yet a recent national poll, echoing many others, found that nearly nine in ten voters across the ideological spectrum say it is extremely important (59 percent) or quite important (29 percent) for the federal government to protect and support national parks, despite federal budget worries. "Likewise," the survey found, "this same spirit of bipartisanship is exemplified by the large majority of voters who say that support for national parks is an issue that can unite people across party lines."

CHAPTER SIX

Destitution Park

Environmental changes confronting the National Park System are
widespread, complex, accelerating, and volatile.
  National Park Service Advisory Board
  Science Committee

Most of the snow-fed source of Bright Angel Creek is a waterfall that
bursts out of caves and fissures high on a wall of the North Rim. It
begins with a roaring vertical drop, levels a bit during the last of the
journey to the bottom of the Canyon three thousand feet down, and
slows. Then you see a fairy tale of a tumbling brook—mercurial fish
and cobbled, shaded meanders. The final run hides in a tight gorge,
passes the cantina and the little cabins and the campground at Phantom
Ranch, and comes out to merge with the Colorado River.

There at its mouth nearly all of the delta created by the creek is bone
dry. The scale of that wide fan of silt and boulders is a hint of the flow
here during a flood. It is also a reminder that a fair-sized fraction of
Bright Angel has been diverted, far up near its genesis springs, into a
sixteen-mile-long pipeline, the sole source of water for most of the 5.5
million visitors to the park each year.

I end a bake-oven day on the cool banks of the gorge, hypnotized by
a sunset skyline of incandescent crags. In the morning I will have company: two Park Service officials to browbeat during an eight-hour hike
back up to the rim.

When you listen to a storytelling park ranger at one of those evening campfire talks or during a guided hike, they typically have little to volunteer about acid rain, biological pollution, budget cutbacks, or the range of other threats to national parks. You're there to enjoy a vacation and, perhaps, to hear something about the Pleiades or the geology or the elk, not jeremiads on the subject of the national park system's decline and its perhaps irretrievable consequences.

Rangers, after all, have to anticipate the sharp edges of our divisive national global-warming debate if they're at, say, Glacier National Park, and they tell us that science research predicts the namesake glaciers will soon be gone. Or, at Grand Canyon, they have to tap-dance through questions about the age of the rocks with visitors who believe the Earth was created six thousand years ago. In the parks much goes unsaid, as science information becomes subsumed by politics or religious belief (and sometimes those are hard to distinguish). With the onset of the Trump era, this politesse has been sorely tried. The administration moved to quash public dissemination of climate-change science data or discussion by the park service and a range of other federal agencies. A "Rogue NPS" social media page sprang up, dispensing resistance messages and "I will not be silenced" bumper stickers that featured the readily identifiable park service flat hat. (It wasn't clear how many actual NPS employees were among the participants.)

As for my two hiking companions, they'll calculate the intent of my questions and observe the usual diplomacies of the government spokesperson. Martha Hahn, the park's top science administrator, and Grand Canyon superintendent David Uberuaga won't leak lurid secrets (both have since retired). With luck, though, I may glimpse something of the changing future of the Canyon through their eyes.

I hope, especially, to learn more about money. One political pastime in recent years has been to demonize federal agencies, starve their budgets, then belittle them for their inability to work effectively—and claim this as evidence that "big government" is wasteful and inept.

Among the three largest federal land-management agencies, the Forest Service is in crisis, for example.

Chronically underfunded in the past, it is now burdened with fighting accelerating waves of forest fires. In 1995 firefighting claimed 16 percent of its budget. Twenty years later that figure rose past 50 percent, swallowing funds that used to be spent on recreation, planning, science research, forest and wildlife management and protection, and the rest of the agency's long to-do list.

"Along with this shift in resources, there has also been a corresponding shift in staff, with a 39 percent reduction in all non-fire personnel," a Forest Service review found. In ten years two out of every three dollars the Forest Service gets from Congress will be spent on fire, it concluded.

The Bureau of Land Management can lay claim to being just as impoverished. It had 10 percent fewer employees in the most recently tallied year than ten years earlier. Corrected for inflation, its annual budget was about 6.5 percent lower in 2015 than in 2005, despite the fact that the agency's responsbilities have increased dramatically, and it brings more money to the federal budget—most of it from oil and gas extraction royalties, than any other—about $13 billion in the most recently audited year. On average the bureau brings in 470 percent more money than its annual budget costs each year.

Sally Jewell, former secretary of the interior, said recently that the Bureau of Land Management is "overextended and underfunded" and can't keep up with the oil and gas drilling boom on public lands. An Associated Press investigation found that four out of ten "high-risk" oil and gas wells near national forests and fragile watersheds, or wells otherwise identified as higher pollution risks, had not been inspected for safety by the bureau. "And we have a major backlog of inspections that people expect us to do on public lands that we are not able to do because we're resource-constrained," Jewell said. Spoken like a true commercial banker—that was on her resume before public service. What she meant was "We have no money."

Then there's the Park Service. Hahn and Uberuaga are grimy and sun-scorched when I meet them the next morning, as they disembark from a week on river rafts—a training trip for park personnel and river guides. Park Service rafters spend broiling months patrolling and ferrying work crews and science teams to some of the remotest places in Grand Canyon.

The spare, measured Hahn wears a backpacker's edition of the gray-and-green Park Service colors and that famous ranger flat hat. Uberuaga, a forty-year veteran of public service, is a grandfather with surplus energy, flushed cheeks, nondescript hiking gear, and a bleached-out orange paisley bandana.

A slender suspension bridge carries the Bright Angel Trail over the Colorado. Slung underneath it is the aluminum pipeline that conveys Bright Angel Creek water down from the North Rim and then back up to the South Rim, a mile above us, when it can. It is a half-century old and the water pressure is seven hundred pounds per square inch. It had to be shut down for repairs just the day before, an event that is at once alarming and routine. It was the seventeenth leak so far this year, and this is only April. Replacing the pipeline could cost as much as $150 million, but there's no firm plan to do so.

Meanwhile, the maintenance guys joke that with all the repairs, they are already replacing it, one foot at a time. It's a suit made of patches. In recent years Phantom Ranch has been evacuated twice because of major breaks that halted the water supply. One prolonged episode threatened to exhaust the whole park's emergency water cache and close it indefinitely. Water was trucked to the South Rim instead, at a cost of $5 million.

"So here we are at Grand Canyon with this billion-dollar operation, all these visitors, and this skinny little fifty-year-old pipe," Uberuaga says as we head up the trail. "How long is that going to last, and are those volumes enough to meet current and future demand? We're filling a thousand rooms with 90 percent occupancy. We're consuming a lot of water. We don't know if one day that valve will shut off and we'll have to regroup."

Despite its mounting challenges and the swelling numbers of park visitors, the Park Service's budget has been up and down, but mostly down. There was one windfall year that provided federal stimulus funds to fend off the impact of the Great Recession, but the Park Service had about six hundred fewer full-time employees than a decade earlier, the last time they were counted.

The pipeline is emblematic of decaying infrastructure at Grand Canyon, a "deferred maintenance" list that totals $371.6 million at this park alone. For the whole park system the figure is $11.5 billion, and about half of it is for road repairs. When you delay those kinds of repairs, the rot accelerates. "That backlog is the best-documented need we have," Bruce Sheaffer, comptroller of the National Park Service, has told me. "Most of it is pretty urgent."

And yet if all the buildings and bridges in national parks and forests fall down, they can be replaced at some future time if we choose, albeit at great expense. But national parks and public lands are suffering from deferred maintenance of another kind, less visible and countable but far more significant and usually irreplaceable: their own natural systems. When species are lost or their populations are in decline, when topsoil is swept away, natural seeps and springs dry up, or invasive species wreak havoc, the damage is often permanent.

Partly because their intrinsic value is incalculably greater than water pipes, electric lines, and asphalt, it is far more difficult to hang a price tag on natural systems or individual species. Maybe impossible. And damage that doesn't immediately affect what humans buy, consume, or profit from sends economists and ecologists down a conceptual maze. What is the number of dollars that we and our governments owe to maintaining the support systems of wolves, salamanders, and Sequoia groves that are under threat? What's the price tag for a Mexican owl, a cougar, a cottonwood grove, or the ecological resilience of Yosemite National Park?

The enormous British Petroleum oil spill in the Gulf of Mexico in 2010 threatened or damaged Everglades National Park and a long list of U.S. marine and terrestrial parks, preserves, and wildlife refuges—as

well as much of the rest of the Caribbean. In its wake federal economists worked hard to try to put a dollar value on the destruction. They
tried to compute "replacement value"—how much it would cost to put
these ecosystems back together or protect others of equal value. Immediately apparent: there's no "replacing" the gulf's natural systems or the
Grand Canyon's if they are compromised by severe enough damage to
prevent their natural recovery. Some smaller pieces might be revived or
replaced—at great expense and with a lot of luck.

Another economic approach tries to determine the value of "ecosystem services" that we humans find useful. It's based on our own economic self-interest. Destroy a wetland and you lose a fishery whose
market worth can be estimated; dam a river and you erase recreational
boating downstream that generates some amount of revenue each year;
cut down a forest and you sacrifice soils to erosion and streams to silta-
tion. You lose carbon storage that could ease global warming.

Ecosystem-services valuations make sense in some ways. They offer
a beguiling certitude and hard-nosed realism about the importance of
protecting the environment. The numbers look firm. Mississippi River
Delta ecosystems provide "at least" $12 billion to $47 billion in annual
benefits to humans and have an asset value as great as $1.3 trillion,
according to one broad-gauge estimate. That's good to bring to a congressional hearing or to some litigation.

Ecosystem-services valuation is deeply repugnant to many natural
scientists, however. Douglas McCauley, a marine biologist, told me that
"if a destroyed mangrove or wetland system in the gulf provided storm
protection and this is the service we want to recover, it is possible that
it could be provided more cheaply and effectively by building levees . . .
and that it could is the potential problem."

McCauley has written that the relationship between markets and
conservation is "ephemeral and illusory. . . . To make ecosystem services
the foundation of our conservation strategies is to imply—intentionally
or otherwise—that nature is only worth conserving when it is, or can
be made, profitable." He asks how we will account for "all of the many

beautiful, evolutionarily old, irreplaceable—and useless—creatures?" The plants and animals, in other words, that are not a visible, countable part of the economy.

Ecosystem-services valuations, the conservation biologist Michael Soulé told me, are "all based on the assumption that economists should make policy decisions. And yet it's a science, or a pseudoscience, that has shown its predictions not to be accurate at all. They can't forecast a depression or a recession. And yet, for some reason, society has vested a great deal of faith in this priesthood of economists, that they know how to determine what's best for society. Trying to place a monetary value on everything is a foolish activity. We don't do it for other important values. For instance, how much is God worth to you? And yet, that's the only thing economists haven't been able to monetize up to this point."

Not long ago The Nature Conservancy hired two polling firms, one with a history of working with conservative causes and candidates, the other with liberals. The conservancy wanted this bipartisan research team to find out what language effectively promotes conservation among people of all political stripes and what words and phrases turn people off.

When their results were in, the pollsters issued a warning. Whether talking to conservatives or liberals: "DO NOT use the term 'ecosystem services.' … Few voters spend time visiting 'ecosystems'—they visit forests, wetlands, rivers, deserts and mountains. And some resist the idea that nature provides 'services' to people—while they acknowledge that people depend upon and benefit from nature, the idea that nature exists to 'serve' them is off-putting to some."

The intrinsic value of species that have evolved to their present form over millions of years and the horror of the idea that unbridled, heedless human activity might extinguish whole life-forms led to the passage of the Endangered Species Act in 1973. There was, at least initially, no pretense of a cost-benefit calculation. That would presuppose putting a monetary value on the survival of a species.

"One would be hard pressed to find a statutory provision whose terms were any plainer," the Supreme Court ruled during an early challenge to the act. "Its very words affirmatively command all federal agencies 'to insure that actions authorized, funded, or carried out by them do not jeopardize the continued existence' of an endangered species or 'result in the destruction or modification of habitat of such species....' This language admits of no exception."

A few years later Congress hedged, or introduced "flexibility," depending on your affinities. It authorized the creation of a cabinet-level committee some referred to as the "God squad," which could exempt some federal actions from the act, even if they threatened an endangered species. It meant that a group of reasonable humans could decide to cost-benefit a species out of existence.

That occurred only a few decades ago, at a time when conservation biologists focused on individual species under the shadow of possible extinction. They spoke of species as essential to the integrity of ecosystems, like rivets in an airplane. One might not seem to matter much, but, imperceptibly, the whole structure weakens as more go missing. Their outlook then looks rose-colored now.

As more rivets popped, the scientists realized that to save a species you have to figure out how to protect a large, sustaining portion of its habitat—and the network of other species essential to the survival of that community. A functional ark needs more than just the plants and animals to bear them into the future. The whole habitat has to be protected.

Soon enough it became apparent that not just a few species or habitats were in jeopardy, but whole classes of animals across broad landscapes, like the millions of acres of the cheatgrass-ravaged Great Basin in the western United States, for example. The politically inert U.S. Fish and Wildlife Service fell chronically behind, far behind, in trying to process hundreds of candidates for its national lists of the "threatened" or "endangered" that were supposed to be under federal protection. So for decades it gave up, to a large extent. (A recent lawsuit seems to have forced its hand. We'll see.)

By now we've learned that much of the mayhem is regional or continental—acid rain and ground-level ozone pollution, for example. In half a lifetime, conservation biology has moved from single species of frogs or flowers all the way to planetwide trouble: polar ozone holes and greenhouse gases.

So I asked Bruce Sheaffer, the Park Service's veteran comptroller, about the size of the deferred maintenance budget for making sure the wolves and weevils and wetlands in the national park system will endure "in perpetuity." When the individual parks send their annual budget requests to Washington, he told me, the "ask" for conservation of natural systems is hard to explain as tangibly as the bad roads and leaky roofs, though of course they make the case as best they can each year.

So what is the price tag for the backlog of unmet environmental needs? "I'm pretty good at coming up with numbers and then using them," Sheaffer said, "and I couldn't even get you to ask your question in a way that I could give you a credible answer. The closest I could come would be to say, well, I know what the parks tell me they need." What they told him a few years ago was that, collectively, they'd need $600 million more. Every year. He doesn't ask for that big picture any more. The case is easier to make in smaller chunks. It's people like Sheaffer, after all, who have to explain the numbers to Congress.

At the pre-Trump time of our Canyon hike, professional travel and training funds had been eliminated there, along with nineteen permanent staff positions, Hahn said. "We're not just falling behind a little bit. We're falling behind a lot," she added. The park's overall budget had been flat for five years. The science and resource management budget declined every year during that period.

For all park operations, including science, "we have so few financial resources, spread so thin," she explained. "The fact is, we don't have a lot of science people in the field. Instead, I have people chasing dollars from universities, from grants, from nonprofit partners. Here at Grand Canyon we don't even have a geologist on staff—is that weird?"

Hahn said she could triple the number of full-time science staff just to meet the most pressing research needs. But the federal government is perceived to be mired in debt, after all; public infrastructure is decrepit, failing; our public health systems are underfunded and demoralized—and that only begins the long, miserable list. So, um, you say we need more scientists at our national parks?

Yes. Public-lands managers can cite spectacular, ingenious, profoundly inspiring conservation projects. They find ways to win support and to squeeze budgets to achieve these successes. When we hear of them, we are reassured that things are okay. Prairie dogs, swift foxes, burrowing owls, and black-footed ferrets are in better shape; wolves have returned to the Tetons after their reintroduction at Yellowstone. The alpine meadows on Mount Rainier, a near-ruin by the 1990s, are far healthier now. The Everglades are, by some measures at least, on a "trajectory of potential restoration"—a lawyered phrase that squints hard to find a hoped-for outcome.

Indeed, such examples can be used to assemble an optimistic, and misleading, mosaic. They are persuasive of what scientists can achieve, but they are far from changing the overall picture. Without a conservation research–funding pipeline in reliable working order over the long term, the parks at times don't even have enough money to monitor their natural resources to determine whether they're thriving or threatened.

When I asked one Grand Canyon scientist about the health of the bighorn sheep herd, for example, he told me there is no way to know, because there hasn't been enough funding to establish how many sheep there are. Similarly, little is known of the park's insect life. Though scarcely investigated by researchers, insects have a major role in the health of Grand Canyon's natural systems. One Northern Arizona University researcher told Uberuaga that he would need the work of two hundred interns just to establish a baseline for insects—to know what's out there—so that future changes and trends could be monitored. "We're going to lose species in the Canyon that we don't even know about," Uberuaga said.

A national advisory committee was recently convened by the Park Service to revisit its resource-management policies in this turbulent era. As one wag noted at the time, the National Park Service churns out plenty of policy studies, full of recommendations. "Given enough time, some of these reports—well, most of them—have been highly effective at warping bookshelves," he wrote. This one is worth attending to, just the same.

"Environmental changes confronting the National Park System are widespread, complex, accelerating, and volatile. These include biodiversity loss, climate change, habitat fragmentation, land use change, groundwater removal, invasive species, overdevelopment, and air, noise, and light pollution.... Parks once isolated in a rural or wildland context are now surrounded by human development. Increasing pressures on public lands—from recreational use to energy development—amplify the importance of protected public lands and waters," one report states. The report concluded that if we want to protect them, they will have to be managed as part of larger regional and continental landscapes.

More frequently now, Grand Canyon and the other parks have to defend against a shifting mix of campaigns to compromise or surrender their resources. "This is an amazing place, and the pressures on it are pretty amazing," Uberuaga said as we moved up the trail. "A lot of people don't see that. They don't see the threat."

Here's the way things work. A hazard materializes: introduced sport fish that undermine the survival of native fish species, uranium mines that could leak radiation into the Colorado, coal smoke and poisonous mercury in the Canyon's airshed, squadrons of rented Harleys overwhelming campground quiet, big lights nearby blotting out the stars at night, or helicopter noise that shatters the solitude on remote trails.

Then, when private interests are involved, there's a controversy over what the park can do to address the issues, or whether they're all that serious. Research has to be undertaken to clarify matters. But the authors of those problems—the miners, hunters, tour operators, or real estate developers—aren't called on to come up with the millions required for the research. The national park, national forest, or some other government entity has to fund it and often does not.

As we've seen, the most imminent threat at the time of this hike involves water. Aside from the Colorado, confined in its channel at the bottom, and rain or snow, no one really knows the sources of the Canyon's water, and they support 80 percent of its diversity of plant and animal species.

We had marched into midday, gaining elevation under the unrelenting sun, but there were breezes, and the general rule is, the higher, the cooler. Behind and below us the Canyon bottom would hit the midnineties today. We turned a corner into some lucky shade, and suddenly a loud and resonating staccato, an eh-eh-eh-eh eh like that of an insistent goat, rang through the stillness.

High up on the shimmering heat of the rock wall across the trail we could see a small niche, a hanging garden full of unexpected greenery, and the source of that startling percussion. It was a male Canyon tree frog, advertising for a mate and warning off rivals—a species that can survive only near seeps and springs.

The mystery of the sources of that subterranean water is not just an obscure academic question. There is deep concern about putting more straws into the natural underground tanks called aquifers around the Canyon region. The severe drought already underway and the specter of climate change add to the necessity of conserving water rather than seeking out more ways to draw down those dwindling reserves.

As we go, Uberuaga polices trash, disperses rock cairns that might mislead hikers onto a false trail, hails muleskinners and family groups and pipeline maintenance workers to ask questions, and surprises a meeting of park planners at a primitive outbuilding to ask more questions. Uberuaga calls himself a "business guy," with an MBA in finance and organizational development. His grandparents were Basque sheepherders, and his parents ran a small grocery store in his native Idaho that often closed for long weekends of camping, hunting, and fishing.

We begin to ascend a sweaty section called the Devil's Corkscrew, and for lack of oxygen I have to turn off the questions between rest stops. Then the trail levels, and we're approaching Indian Garden, a shady oasis halfway to the rim. The conversation turns to possible sources of money for research and for everything else that sustains this impoverished glory.

I strike up a conversation with three Belgian college students touring the United States on a slender budget. An amazing 38 percent of visitors to Grand Canyon are foreign citizens. This is partly the result of a federally supported marketing campaign to bring in more foreign visitors and to reap the economic benefits. For millions of them, and for Americans too, national parks are the draw. So they are induced to come and spend money, en route, in Phoenix or Las Vegas or Tusayan. Everyone does well, except the destitute parks.

National park visitors put an estimated $15.7 billion into the cash registers of private businesses in local gateway regions, in just one recent year. The money directly supports nearly 174,000 jobs and a $5 billion private-business payroll. Many of us want to dip our buckets into that revenue river. When budget fights in Washington have shut the parks down, the outcry from the business community is, understandably, instantaneous.

"What if Grand Canyon charged extra for people from other countries?" I asked the students. "It would actually be quite okay," one said, "especially for the campsites. We paid only six dollars a person. In Europe you won't find any campsite for that. In Belgium I'd say it would be twenty dollars per person" (and not at the scenic equivalent of Grand Canyon, I would guess). "I found it amazing that the prices for the parks are so low. It's crazy," another student put in. All three told us that they'd willingly pay much more, both for the park entrance fee and the campground fee, before they'd consider changing plans. That alone could bring in tens of millions each year.

It's not just the foreign contingent that could afford more. When Americans arrive at Grand Canyon, we fork over an absurd thirty dollars per carload to be in the park for a full seven days. It took years of campaigning just to get the fee raised to that level. Tour bus operators pay eight dollars per passenger. Helicopter air-tour companies are charged only twenty-five dollars per flyover, sometimes less, often nothing—no matter how many passengers are in the aircraft. More than one hundred thousand overflights carry nearly half a million passengers each year.

In Kenya, by contrast, each national park visitor pays per twenty-four-hour period, plus an issuing fee, plus a fee for a car, and the entry and exit are time-stamped. Ecuador's Galápagos National Park costs a hundred dollars per person to enter, unless you're Ecuadoran. Argentina's national parks charge a per-person entry fee higher than what Grand Canyon charges per carload. Uberuaga said that whenever he asks them, visitors say, "You ought to charge us more! We're okay with that!"

On a day when I decided to pay attention to hardware instead of nature I watched the lineup that pulsed through the Canyon's south entrance, and I toured a couple of parking lots. Recreational vehicle rentals are popular, it looks like. They begin at a hundred bucks a day, plus a mileage fee. But most people drive their own elephantine Unitys, Westwinds, Sunseekers, Leprechauns, or Winnebagos. Prices for RVs like these begin around $50,000 and head north rather briskly toward $200,000.

I looked hard for examples of cars that were down on their luck—the kind that community college students might drive or folks who can't afford much in the way of travel expenses. Almost none. Far fewer than the number of Escalades, Mercedes, and girthy SUVs. No wonder so many visitors say they're willing to pay more to enjoy national parks.

But those are self-serving guesses, I admit. And it's a mistake to be blithe about spending other folks' money. In fact, no one knows much about the ages or the income levels of park visitors. Without great delay and difficulty, administrators of U.S. public lands are not allowed to pose routine marketing questions of visitors, like "How high would entrance fees have to be before you'd stay away?" That "demand elasticity" is easy to model for both foreign and U.S. visitors, Bruce Sheaffer told me, but it's politically impossible.

"You'll get no argument out of me that fees shouldn't be considerably higher than they are," he said. "They absolutely should be higher than they are." But the political impasses that prevent more federal support for national parks also shut down increases in user fees. Uberuaga said that a fee increase is a nonstarter: "The lobbyists for the American Association of Retired Persons, the tour bus industry, all those guys know what they want."

An estimated 40 percent of Grand Canyon visitors get in for next to nothing. The figure includes active military and veterans who are allowed free entry, as of course they should be. A far higher number are beneficiaries of the senior citizen bonanza: if you are sixty-two or older, for ten bucks you have a lifetime pass to all national parks and federal recreation areas. Campsites are an unreal deal too: "You pull that RV in," Uberuaga advises. "It's the cheapest campground in northern Arizona. And then you get half off because you have your senior card."

Any attempt to raise the cost of that plastic card is paralyzed, Sheaffer said. "The AARP [American Association of Retired Persons] stops us in our tracks. The politics are stunningly difficult to overcome." I contacted the AARP to ask whether most older Americans might not be happy to shell out their fair share to get into the parks. "AARP does not lobby Congress on park fees," was the reply. So I invited the vice president in charge of answering such questions to consider whether the association would "support an initiative to abolish senior citizen passes and discounts at U.S. national parks in view of the parks' financial difficulties." Still waiting for a reply.

Senior passes may be good politics—I doubt it—but I figure that merely getting old shouldn't earn me a lower-than-dirt-cheap lifetime admission. (I'm seventy.) As for the oft-heard homily that "the parks are for the people," and people of modest means can't afford a fee increase, I salute that in good will. I propose a big improvement: anyone who shows up at a national park with food stamps gets in free.

Fee levels for users do not derive from their needs, nor those of Grand Canyon or Utah's Grand Staircase National Monument or Virginia's George Washington National Forest. They aren't the result of the willingness-to-pay of visitors or of the public interest in public lands. They're just the wrong kind of politics: lobbying, connections, campaign contributions, and antigovernment cant.

One Park Service preoccupation about fees is astute in some ways, but perhaps more so in an earlier time. If you throttle back on the number of visitors by limiting them or by raising prices, the thinking goes, you erode political support.

This is a live concern, but plenty of research shows that people often place a higher value on what they pay a higher price for, all else held constant. It's called the "Grey Goose effect" by economists, after the namesake vodka that is, except for its brand identity and very high price, pretty much like any other vodka.

The 2016 budget request for the whole park system was $4.2 billion. Congress approved $3.4 billion instead. Visitation to all national park units in 2014 jumped 5 percent over the prior year, setting a new record. Another 5 percent in 2015, another new record. The centenary year for the system, 2016, set still another new record.

There are many ways to reprice the parks on an experimental basis, and plenty of trade-offs to be calculated. Except for political inertia, parks could easily be allowed to raise entrance fees and then gauge the visitor response. Consider this oversimple plan: Many national park units don't charge an entry fee at all and some, for practical and legal reasons, can't. But for entry to the 135 park units that do, let's charge every park visitor of any age—except poor people, active military, and veterans—twenty dollars per day. (Disneyland: $100+ per person for one day.)

Drop the number of annual visitors by half, just as a fudge factor, because you think that's how many will stay away and play video games instead or because of other complications. Estimate, preposterously, that each visitor stays only a day. The Park Service would take in about $1.2 billion a year—at least a billion more than entrance fees total now. And in this happy scenario the usual annual congressional budget allocation would of course stay at current levels too.

Some fraction of that revenue could implement a rolling management plan for climate change on all public lands. "Every young person joining the Park Service now, their entire career will be consumed by climate change and responding to it," according to Gary Machlis, the science adviser to the director of the national park system. "We need to train them and prepare them for those complexities. We have to take the time to do that." And we have to pay the money.

CHAPTER SEVEN

Air and Uncle John

Lots of people think of pristine places as refuges from fouled air, but fouled air is transported over great distances.... When you look at bad stuff that you breathe, whether it's pollutants or cigarette smoke ... a little bit harms you a little bit. A lot harms you a lot.

> Dr. Norman Edelman, chief medical officer,
> American Lung Association

The first view of the old building poised at the edge of the South Rim at Yavapai Point always pleases. Its scale is modest and its roof, mesa-flat. Rough Kaibab limestone walls and exposed Ponderosa beams evoke the colors and ledges of the Great Canyon beyond and below.

Now the park's Geology Museum, this was once known as the Yavapai Observation Station. It had a special function when it opened in 1928, and it was sited here after keen deliberation by a blue-ribbon team of researchers—a bunch of charming greybeards whose photos are now part of an exhibit inside.

Despite the declamatory style of their day, even in their official statements these worthies seem besotted with the Canyon and its endless trove of scientific revelations and eager to share them. So the building was designed to focus outward on what one early chief naturalist called "the park story—the whole damn story." In other words, this place offered the finest vantage, they decided, for interpreting the geology of the Canyon, that two-billion-year storybook, for visitors.

Anyone who has spent more than ten minutes at the rim can argue the merits of his or her own most-preferred spot. But here a comprehensive 180-degree view of temples, buttes, and far-off gorges and rockslides are well framed by the big windows. There's even a glimpse— only rarely offered along the thirteen-mile, strollable South Rim edge trail—of the shifting hues of the Colorado far below.

Suppose for a moment, though, that the Observation Station's view has not been well cared for. Maybe the exhalations of tens of thousands of visitors each year and the sticky fingers of phalanxes of schoolkids have left heavy smears all over the inside of those essential windows. Add—why not?—an eager gang of cigar enthusiasts. On the outside of the windows, let's say, a streaky layer of dust and cobwebs has accumulated on the glass. So if all that were allowed to happen, you would strain to see the Canyon through a shroud of grit and greasy haze.

That's a far-fetched scenario, but as it happens, its equivalent in visual terms comes close to our reality on many days. The problem is out there in the Canyon air, though, not here inside. With a little squinting we can make out its sources, impacts, and possible solutions.

The Grand Canyon air-pollution story differs in detail but not in kind from those of the whole national park system and all the rest of the national landscape. We can't solve the problem just for the parks, any more than we might clear the air in one corner of your living room if some of your guests are smoking in the middle of it. Air pollution is a general affliction. It poisons water and soils and has serious consequences for human health, and natural systems. The loss of visibility is only its most immediate symptom.

Part of this story is told by a discreetly camouflaged network of instruments—filters, sniffers, detectors, gauges—designed to assess what's in the air. There are fifty of them altogether in several locations at Grand Canyon, and they were tended, during one of my visits, by a Park Service air-quality specialist named Shannon Reed.

On a hazy spring morning she took me along for her daily round of data gathering, which ended at the Geology Museum. Within the Canyon the gray veil thickened. She pointed out a nondescript padlocked white box at the very edge of the rim, under the shade of a Gambel oak. It houses a transmissometer, an instrument designed to track changes in visibility. It looks like a telescope whose barrel is aimed at a downward angle, toward a companion instrument that shoots a beam of light upward at us from near Phantom Ranch, at the bottom of the Canyon.

I looked into the eyepiece and through two miles of Canyon air and saw, to Shannon's relief, a faint answering pinprick of light from down by the river. When that faraway light bulb burns out, she has to hike the ten-mile trail down to replace it.

"If it was 100 percent, if this was a beautiful day, you would get a visual range of about 240 miles," she said. The transmissometer compares the light it should receive on a perfect day with what it's really getting. It measures the scattering and absorption of the light. Then she can calculate the loss in visual range. On this day, the thirteen thousand or so people here at the park have lost quite a lot. We can see out just 61 miles—about a quarter of the distance that natural conditions would afford on the clearest of days. The atmosphere on some of those no-pollution days would not be perfectly clear either, though. There's a varying amount of natural stuff out in the air at all times, but the natural visual range here averages a still-gorgeous 170 miles.

Let's nail down the arithmetic. It tells us that on this particular day, we've lost about two-thirds of the visual range of the *average day* we'd see under natural, pollution-free conditions. If we change the comparison, it looks this way: we've lost three-fourths of the visual range of the *clearest day* we'd see under natural conditions. Grand Canyon's air is impaired by some level of human-made pollution 90 percent of the time. Never mind just the day Shannon and I were at Yavapai Point. On average, pollution reduces the Grand Canyon's natural visual range by more than 30 percent.

Our visual perceptions aren't only about distance. The Canyon view we celebrate is enormous in scale, but not really so far away. The North Rim across from here at Yavapai Point, for example, is only ten miles off. Pollution degrades views both near and far, though. In the words of a National Research Council study of air pollution in the parks, their haze "reduces contrast, washes out colors, and renders distant landscape features indistinct or invisible."

If we are lucky enough to be here at Grand Canyon at all the visit will be brief for most of us, and just once or twice in a lifetime. The view is nearly all of the experience. Gazing down-Canyon, John Muir once extolled "the impression of wild, primeval beauty and power one receives in merely gazing from its brink … the view down the gulf of color and over the rim of its wonderful wall."

So what is in that scene-stealing haze? There are natural sources and anthropogenic ones—that somewhat syncopated term means "human-caused." Even on an exceptionally clear day, oxygen and nitrogen in the air deflect the light waves that bring a distant image to our eyes. That's natural—it is called "blue sky scatter." The list of natural sources also includes wind-blown dust, chemical haze that arises from vegetation, smoke from the increasing number of wildfires in the region, and, surprisingly, even a measurable smidgen of sea salt.

An occasional source of haze is the smoke from "controlled burn" fires, set purposely by the Park Service, the U.S. Forest Service, and other agencies to try to clear woodlands in the Canyon region of a century's accumulation of scrub. But year in and year out, the tall stacks of coal-burning electric power plants and the tailpipes of diesel and gas engines near and far—as far away as, say, Los Angeles or Seattle—account for most of the air pollution here.

Our national push for better air quality began in earnest a century ago, and the spur was toxic particles and fumes from smelters in a place called Copper Basin, Tennessee. They were killing forests and orchards and making people sick across the border in Georgia. Tennessee refused to rein in the copper companies and disputed Georgia's right to interfere. A U.S. Supreme Court decision penned by the clearly indignant chief justice, Oliver Wendell Holmes, settled the issue in Georgia's favor:

"It is a fair and reasonable demand on the part of a sovereign [state] that the air over its territory should not be polluted on a great scale by sulphurous acid gas, that the forests on its mountains ... should not be further destroyed or threatened by the acts of persons beyond its control," Holmes's 1907 ruling explained. As we know, only a few years later Congress created the National Park Service "to conserve the scenery [in the parks] ... and to provide for the enjoyment of the same in such manner and by such means as will leave them unimpaired for the enjoyment of future generations."

Most of the data on the Canyon's air quality was collected after the passage of national clean-air legislation in the 1960s. Those laws are part of a slow, faltering progress. The view from the rim now measures its limited success. You could look at this as a long campaign to build a protective fence around the clean air of the natural places in the United States that we most value. But because there is no such thing as fences that keep out dirty air, the effect of the rules went much further, potentially: they required that the sources of any pollution that reaches pristine areas was to be tracked down and throttled back—Oliver Wendell Holmes's century-old ruling, revived. That would mean much cleaner air in a lot of other places too, a sort of virtuous tail-wagging-the-dog.

These moves over past decades have cleared the air to a degree, but they are implemented very slowly. Acknowledging the expense and technical hurdles to be overcome in reaching the goal of "natural air quality conditions," one major deadline was set at the year 2064. "Reasonable progress" toward the goal had to be demonstrated to the Environmental Protection Agency (EPA) by each state, however. That progress, or lack of it, has been discussed and litigated ever since.

You may sense, correctly, that there has been strong reluctance in some quarters to keep up the momentum toward clean-air goals. In 2002 Eric Schaeffer, head of the EPA's Office of Regulatory Enforcement, resigned to protest political interference with federal clean-air regulation and enforcement. His anguished resignation letter lamented "a White House that seems determined to weaken the rules we are trying to enforce. It is hard to know which is worse, the endless delay or the repeated leaks by energy industry lobbyists."

By 2007, as the deadline for antipollution plans for protected areas such as Grand Canyon came and went, almost none of the states, including Arizona, had come up with them. It was the Environmental Protection Agency's job to enforce compliance, but the agency let that target year slip past, more or less without a murmur. A coalition of environmental groups sued the EPA to force it to follow the law. Finally, years later, a federal judge ruled against the agency and the laggard states and imposed a new set of deadlines.

So we can relax, now that the lawsuit is settled? "Completely wrong," attorney Stephanie Kodish of the National Parks Conservation Association told me at the time. "The lawsuit and the settlement only require that states and the EPA make plans. It does not at all speak to the substance of those plans. We have no guarantees."

Smoke from the western wildfire epidemic of recent years clouds the issue. Just the same, some of those clean-up plans that the states have submitted to the EPA may not strike you as good-faith efforts. Here's part of my conversation with Don Shepherd, an environmental engineer with the Air Resources Division of the Park Service:

NASH: You mean, then, that a series of deals have been cut that have pushed the practical deadline back forever?

SHEPHERD: Well … in many cases, maybe most cases, they're looking maybe a hundred or more years beyond 2064 and then … Arizona is looking at something like eight thousand years beyond 2064. Essentially, they are nowhere close to making the amount of progress that they need to, and nobody is really holding their feet to the fire on that.

NASH: If somebody were to, would it be the EPA?

SHEPHERD: [Laughs.]

NASH: I mean, are they the agency that's supposed to do that enforcement?

SHEPHERD: Supposed to, yes.

NASH: In fact, nobody was even asking for those plans until some litigation transpired, right?

SHEPHERD: Right, yeah.

NASH: Eight thousand years, how can they propose that with a straight face?

SHEPHERD: That's a really good question. I've wondered about that too. Even the state plans that go a few hundred years beyond 2064, it seems like that would be kind of hard to propose with a straight face too.

NASH: So it's the case that the EPA, which is often maligned for the ferocity of its interference with a market economy, is just winking at that 2064 deadline?

SHEPHERD: I think winking might be—you put it more aggressively than what they're really doing.

NASH: That eye is blind.

SHEPHERD: Pretty much.

The take-home: there is a bulky catalog of ways to try to make sure that promises are not kept. Players like the U.S. Chamber of Commerce, the National Association of Manufacturers, and state-level interest groups oppose air quality regulation with routine, bitter tenacity. State governments know how to stall on their behalf. When political winds blow, federal agencies bend. The Trump-appointed head of the Environmental Protection Agency, Scott Pruitt, has not only rolled back the Obama-era Clean Power Plan, but agrees with those who say the agency itself should be abolished. The administration proposed a 31 percent budget cut for the EPA for 2018.

At the Canyon, visibility has improved marginally by some measures, compared to the early 2000s. Whether we're deciding to go backward or forward or to stand still on air pollution, though, it would be good to know what the trade-offs are, not just for Grand Canyon or our other public lands—some of which suffer far worse pollution—but for the well-being of us.

I think of the most visible part of this debate as the "Uncle John" problem. Whenever environmental fixes are under discussion, I have a friend who is careful to focus on the likely human costs as well as the benefits: "What about Uncle John? He needs a job!"

He is correct to ask. It's the question posed when someone talks about ranchers getting cattle off public lands; about helicopter pilots flying noisy air tours; about sprawling real estate developments at the edge of national parks and forests, and about agencies slowing international trade with better inspections to keep out invasive species.

But the "jobs versus the environment" discussion is often a false dilemma, expensively promoted by larger, cloaked claimants. The Clean Air and Clean Water Acts, the regulation of drugs and automobile safety, miles-per-gallon auto standards, workplace safety, and child labor laws were all forecast as doom for the economy. They have not been. The job question can be just a fig leaf to camouflage more narrow economic interests.

These false alarms also routinely inflate one factor in the equation—economic impact—and ignore others. Plainly, we could follow the example of, say, China and contrive the employment of hundreds of thousands in the construction of firetrap housing, exploding cars, or poisoned food and drink. We choose not to, because of the consequences. We assume, then, that economic health, human health, and environmental health are not incompatible. Yet in the case of air quality at Grand Canyon, the jobs question cannot just be wished away.

Let's try to assemble the pros and cons into a coherent picture as we all, Uncle John included, figure out how to transition away from fossil fuels and dirty air. The trade-offs are on display at the Navajo Generating Station (NGS), on tribal land only twelve miles from the boundary of

Grand Canyon National Park. A coal-burning power plant, it is the largest in the West. You can see its three colossal stacks—among the tallest structures in Arizona—from miles away, in looming isolation on the desert floor. The NGS is unusual in scale, and its fate was uncertain as of 2017, but it is not an isolated example. More than a thousand such coal burners generate power around the United States—many of them delivering fouled air across national parks and other public lands, not to mention urban areas. Its particulars, then, give us a far broader image of pollution politics.

After a long battle the plant's owners and operators, mostly government entities, were forced by the EPA to install scrubbers to trap some of the worst of its pollution—sulfur emissions—as the past century ended. But the nitrogen oxide from those stacks still accounts for most of the haze that afflicts the Grand Canyon, according to EPA modeling (which is disputed by the plant operators). The plant also generates an important fraction of the haze in at least ten other natural areas in every direction of the surrounding region—many of the most beautiful parks in the Southwest and in the nation.

The NGS came online in the mid 1970s. It is the roaring heart of an improbable, exuberant, and ultimately destructive feat of post–World War II engineering, emblematic of the era and its ambitions. The bipartisan boosters in Arizona and in Congress who enabled its creation were warned by federal scientists in advance that it would prove to be a monster polluter. But it has other uses.

Each day twenty-two thousand tons of coal are stripped from the Kayenta mine on Black Mesa, on the Navajo and Hopi reservations. It is hauled to the power plant by three trains over a 78-mile-long dedicated electric rail line. Burning that coal generates about fifteen million tons of climate-changing carbon dioxide a year.

The water it needs to make steam—an annual nine billion gallons—comes from nearby Lake Powell, behind Glen Canyon Dam. A quarter of the plant's electric energy output is sent over a 300-mile transmission line to power giant pumps much farther down the Colorado. Those pumps are the largest single user of electricity in the state, and they suck about 520 billion gallons of water from the river each year. Then they push it through an aqueduct 2,900 feet up and across the Hieroglyphic Mountains and the Superstition Mountains and 336 miles south, to supply a third of the water consumed by thirsty Tucson and Phoenix.

The entire epic contraption has radically altered much of the region's economic and social history over the past four decades. It supplies subsidized power to Arizona, Nevada, and California and moves most of its subsidized water to the deserts of southern Arizona, metropolized with the help of this imported water supply. The rest goes to tribal and private agricultural lands along the way. The plant is woven into a knot of legal, political, and economic arrangements that go back half a century.

If the plant were closed, the lights would not wink out in southwestern households, though. The NGS is part of a network of energy suppliers across the West, and federal reports say there's a surplus in the region: "Shutting down Navajo GS would change the regional flow of power," a report by the federal National Renewable Energy Laboratory states. But replacement power would come from natural gas plants in Nevada, Arizona, and Southern California that already exist and have spare capacity. The analysis also predicts that energy costs at the big pumping stations that push water to southern Arizona may rise if the NGS were closed. You could look at the plant's pollution, then, as a grim subsidy for cheap water and power in Phoenix and Tucson.

The Uncle John problem is especially pressing here, however. The Navajo tribe's unemployment rate is estimated at 40 to 50 percent. The plant and the coal mine together contribute more than $150 million a year to the Navajo and Hopi tribal economies, including 850 jobs that pay about twice the local average. They also pay tens of millions in royalties and other fees to the tribes and several hundred thousand dollars a year in scholarship funds and property taxes that go mostly to reservation schools.

Judy Moffett, a Navajo, is one of the engineers who drive the electric locomotives that pull mile-long trains of coal to the NGS. She has worked there for more than thirty years. She explains, in the polished prose of a company press release, that she's grateful for the job, because it has allowed her to fulfill her dreams for herself and for her family. "A lot of people would be out of work if it wasn't for NGS," she says. "A lot of kids wouldn't have gone through college if it wasn't for NGS. A lot of older folks wouldn't have a better life if it wasn't for their kids working at NGS and looking after them, you know, making life a little easier for them."

Somewhere within the labyrinthine power plant I spoke with Jarbison Littlesunday, a Navajo who has presided for fourteen years over a busy control room. During my visit he fielded intercom calls, attended to alarm bells and horns, and tracked a bank of meters and video readouts signaling boiler pressure, water pressure, turbine revolutions, and combustion temperatures. He trained for five years before promotion to this work. "The money I make here actually goes back home to my family, who live in the area where I'm from," he told me. He knows the pollution controversy well: "Everyone is watching this plant," he said. Now there are two cards on the trade-offs table: visibility and jobs.

The third is health. Unfortunately, dirty air doesn't just affect the view; it also affects human lungs and hearts—and global climate. One component of the Canyon's air pollution is ozone, sometimes called photochemical smog. It is an unstable form of oxygen that combines relatively easily with the chemistry of other materials, including living organisms, in the process called oxidation. It's confusing at first, but six to thirty miles up in the stratosphere, a layer of ozone is essential to protect the Earth from incoming ultraviolet radiation. "Holes" in this ozone shield, caused by synthetic chemicals, are a continuing concern—about skin cancer, for instance—despite some international success in controlling them.

Ozone also occurs naturally down here where we live, but human activities now add a far larger quantity of it to the air, and not just in cities. Ozone is created when two kinds of pollutants are cooked by the sun. One of them, nitrogen oxide, comes mostly from power plants, industrial boilers, and internal combustion engines. The NGS, for example, exhausts some fourteen thousand tons of nitrogen oxide into the sky each year. That affects human health as well as visibility.

Ozone pollution inhibits growth in a long list of agricultural crops. It makes at least two species of western pines grow more slowly or die. Closer to home: in Grand Canyon, in many national parks in every region, and across huge areas of North America, ozone occurs at levels that can affect your lungs.

On public lands and the regions that surround them, rapid oil and gas development have also worsened ozone pollution over broad tracts of the rural West. The supposed imperatives of fossil fuel energy development, and their politics, often paralyze enforcement of clean-air laws. Around Theodore Roosevelt National Park in North Dakota and Carlsbad Caverns National Park in southern New Mexico, "the oil and gas development is just overwhelming any other emission reductions that are taking place," the NPS environmental engineer Don Shepherd told me. "It may be a little mind-boggling, but of all places southwestern Wyoming is currently exceeding allowable ozone limits for the federal health standard and so is the area around Dinosaur National Monument in Utah. The only possible causes for very high ozone in those remote areas are the oil and gas emissions."

Odorless, colorless, but far from harmless, ozone causes respiratory symptoms and exacerbates disease, according to Norman Edelman, chief medical officer of the American Lung Association. It's something like a sunburn that irritates and reddens lung tissue. It's harder on children and the elderly, and worse if you are exercising—going on a hike, for example. Multiple exposures over time, some research shows, can affect the elasticity of the lungs, just as sunburns do your skin. It can reduce the immune system's ability to fight off bacterial infections in the respiratory system. "You know, about twenty million Americans have asthma," Edelman told me. "When they're exposed to ozone, which is a highly irritant gas, there's ongoing inflammation in their lungs and airways. For them, the inflammation increases, and they're susceptible to an asthma attack. Now one of the things you have to understand is that anything that causes inflammation of the airways has a synergistic effect. So if your airways are slightly narrow because you have asthma, and they almost always are, and then they're narrowed more because of the inflammation of ozone, and then if you hike or do other exercise, you're much more likely to get an asthma attack.

"Some people are not asthmatic," Edelman said, "but they are especially sensitive to ozone anyway—maybe 10 percent of otherwise healthy adults. There's evidence that chronic exposure to unhealthy levels of ozone affects long-term lung function and that children wind up with poorer lung function if they're in high-ozone areas," he added.

Walking at Grand Canyon or another national park on a bad ozone day you may find it more difficult to draw deep breaths than you normally would. Breathing may begin to feel uncomfortable, faster, and more shallow. "More recent data actually shows a relationship between ozone and death," Edelman said. "And now it's beginning to look like it has an impact on cardiovascular health as well. The strongest association so far is between ozone and the blood vessels of the heart."

During the most recently monitored year, the hourly readings at the Abyss, a scenic overlook in the Canyon, registered ozone levels deemed unsafe by the American Lung Association 277 times. Some of those are multiple readings in one day, so most days are okay. Typically, visitors spend only a short time in the park. It's not that a single visit there is a long-term health risk, though a bad day can be a short-term problem for asthmatics and other ozone-sensitive people. Things could be worse—and they definitely are, especially in cities and in eastern national parks and forests. So the question is this: if so much pollution regularly settles on Grand Canyon, what does the rest of the United States look like from inside your lungs?

"Pollutants are synergistic," Edelman said. "Lots of people think of pristine places as refuges from fouled air, but fouled air is transported over great distances. Is there a threshold effect, a level below which it's safe? When you look at bad stuff that you breathe, whether it's pollutants or cigarette smoke, most of the data suggests there is no threshold. I mean, a little bit harms you a little bit. A lot harms you a lot."

During the decades-long negotiations over what to do about the NGS, the National Parks Conservation Association commissioned George Thurston, professor of environmental medicine at New York University, to try to calculate the human toll from nitrogen-oxide particle pollution from the NGS. In the state of Arizona, it results in as many as five premature deaths each year as well as several heart attacks and cases of chronic bronchitis, he found.

His report takes an additional step. The EPA places a dollar value on various health problems and death, based on people's "willingness-to-pay" to avoid these consequences. So how much would you pay to avoid death, or chronic bronchitis, or some other illness caused by air pollution? The method may sound perverse, and it's certainly controversial, but it's all we have if we decide that money is the meaningful measure of value in this kind of environmental poker game.

Thurston concentrated on the health effects of particle pollution from nitrogen oxide and opted not even to include the "sunburned lungs" and other ozone-related kinds of health problems in his calculations. That means, he told me, that his figures are quite conservative. Without a nitrogen-oxide fix, the plant will continue to "exacerbate the substantial, and irreparable, harms to public health that have already been incurred to date," his research found. Thurston's figures, using the EPA's price tags on life and health, show that if the EPA does not enforce more stringent nitrogen-oxide emission standards at the NGS, the health impact will total $140 million to $350 million over the next ten years, including up to fifty premature deaths.

The NGS also sent fifteen million tons of carbon dioxide into the atmosphere in the most recent audited year. Other coal-powered plants in this sector of the Southwest double down on that figure. The gas rises through the sky and stays, some of it for centuries and about a quarter of it essentially forever—a hundred thousand years or so. It becomes part of the global acceleration of climate change caused by greenhouse gases. If carbon dioxide is figurative "dirt," then Navajo ranks third among the dirtiest power plants in the United States.

Obama-era federal regulations were underway to impose controls on carbon dioxide, and then were targeted for rollback by President Trump, industry lobbyists, and an eager majority in Congress.

Finally, the big stacks of the NGS cast several hundred pounds of mercury into the skies each year, and its coal-burning cousins in the neighborhood add more. A warning has been issued by authorities about high levels of mercury in striped bass at Lake Powell, six miles from the NGS. Newer research suggests that mercury has also polluted the Colorado River inside Grand Canyon.

This study found indications of mercury concentrations in insects, snails, and organic matter that fish feed on. It often exceeded risk thresholds for fish, wildlife, and humans. The source of the mercury contamination above and below the dam is not clear, but additional studies are planned.

Exposure to hazardous levels of mercury over time can cause a wide range of neurological damage, including vision, speech, hearing, coordination, and sensory impairment in humans. It is also linked to lower reproductive success, growth, and survival of fish and wildlife.

Visibility isn't just aesthetics, then. It's a proxy for health. And national parks and public lands are a proxy for concerns about pollution everywhere else we live, with air pollution causing thousands of deaths, heart disease, and other ailments across the United States, according to a long series of research studies. In a sense, then, we all live at Grand Canyon.

So air-quality trade-offs—visibility, health, jobs, and the politics of artificially cheap water and power—have imposed gridlock here for decades. During my visit to the NGS, I sat with a couple of public relations people and the plant's environmental and operations manager, Paul Ostapuk. They spoke of their concern for an orderly rather than an abrupt transition away from fossil fuels so that hard-pressed tribal families and tribal governments can adapt over time.

"We don't want to just walk away," I was told. "There is going to be change, but we feel like we need to get there in a way that's responsible to our customers." The timetable the operators had in mind, they said, is to take another three decades to step entirely away from coal. Not long after, events would make that professed concern for tribal employees a curious paradox.

I asked whether research was underway on the feasibility of solar-energy conversion, to keep jobs on the reservations. "I guess we'll leave that to folks like you to dig into," Ostapuk laughed. I get it. Their campaign was to keep the power plant they already have.

So, fair enough—someone else has to figure a pathway out of the fossil fuel fog. The Arizona desert's a sunny place—"fantastic solar resources," as a National Renewable Energy Laboratory analyst told me—with plenty of flat, vacant land. How about we just pull down that megasaurus coal plant and slap up a big bright field of solar panels instead?

It's not so tidy as that, but several groups have made a start. One obstacle is that solar power can be stored only on a modest scale. Without storage a solar plant's output can't be adjusted to track demand, as fossil fuel plants can. At night or when the clouds come, solar goes idle. During those times demand across the grid has to be met from other sources of power: hydro, wind, nuclear, somewhere else's sunshine, or fossil fuels.

On the other hand, solar generates maximum power during the season and the time of day when a lot of it is needed, to meet demand created by millions of energy-sucking air conditioners. And the sun is free, so once the system is in operation it costs very little for the resulting energy compared to coal, oil, or natural gas.

In any case, swapping out coal for sun in this spot presumes that generating electricity here is a necessity. But as noted earlier, it isn't. The foreground issue isn't losing energy supply. It's losing cheap water and power in Phoenix and Tucson, and good jobs and tribal income where they're desperately needed.

The National Renewable Energy Laboratory has published a preliminary study of solar, wind, and geothermal energy sources that could supply jobs and heat to run the big steam turbines at the NGS. It would be expensive and complex but feasible. "Expensive" is a relative term, and it depends on how you calculate the negatives of air, water, and atmospheric pollution.

Other federal studies, roughly adapted by a couple of their coauthors at my request, suggest that to replace all of the NGS's output, they'd need to blanket about a hundred square miles with solar panels—by coincidence that's about the same area as the Black Mesa coal mine leases that supply the power plant now—at a cost of something like a whopping $10 billion to $12 billion. Those initial costs and a payback schedule are other entries on the trade-off ledgers, just as they were when the NGS was built, decades ago, with $3.2 billion in government cash (in today's dollars).

Leaders have emerged within both the Hopi and Navajo tribes to push for closing the coal mine and the power plant and to replace them with solar and wind energy. Theirs is a minority view for now. "Coal mining is a short-term crutch," Marshall Johnson, the founder of one such group among the Navajo, has written. "For short-term royalties and a few jobs, our government looks the other way while mining activities take an irreparable toll on the health of the people and the environment of Black Mesa." There is another way, he argues. "We can transition to long-term, sustainable economic development opportunities like clean-energy manufacturing and installation."

The group commissioned a report by a Massachusetts consulting firm, which concluded that hundreds of temporary and permanent jobs would be created with wind and solar farms, as well as millions of dollars in annual royalty payments to tribal authorities—though not enough to replicate the NGS's current cash infusions.

Former Hopi tribal council member Vernon Masayesva directs a group with similar aims: a solar power and natural gas plant to be built on the reservation. "It's entirely feasible to build that plant right now," he told me. "A thousand-megawatt plant, about half the size of the Navajo Generating Station, would generate 2,500 construction jobs for at least three years, plus 300 permanent jobs."

That does not resolve the employment issue as certainly as just keeping the dirty, old plant in operation, with all its liabilities, indefinitely. One of the reasons the federal government orchestrated its construction in the first place was to provide an economic boost for the Navajo and Hopi tribes. If that is still a national goal, there are now much cleaner ways to pursue it.

Abruptly, however, the NGS operators pivoted. They announced in 2017 that they planned to shut it down by the end of 2019. Power plants in other places along the distribution grid that use natural gas could supply electric power to customers more cheaply.

Alternatives to keep NGS open were immediately proposed. Perhaps the coal-fired campaign rhetoric of the Trump administration would somehow be made good on, despite the economics advanced by the NGS operators. Whatever becomes of this plant, a legion of others including several in the Canyon region will continue to burn coal until some other factor intervenes. Perhaps it will be the long-term health of the atmosphere, the landscape, and their wild and human inhabitants. Or, perhaps, the deciding factor will continue to be the short-term economic health of a power plant's owners.

Natural gas is said to emit only about half of the carbon dioxide that coal does when burned. That is a major improvement, though the figure is disputed. Some natural-gas extraction methods, notably "fracking," loose another powerful greenhouse gas, methane, into the atmosphere. Gas mining also brings a host of other destructive environmental problems to the landscape.

Non-fossil-fuel energy sources—the "renewables," chiefly wind and solar power—are far more benign. But I've been one of those who greet renewable-energy advocates with a furrowed brow when they put forward plans for conversion. I've read and heard how convulsive the switch away from fossil fuels will be for the infrastructure, the economic system, and employment. A couple of years ago, however, the citizens of Germany, the strongest economy in Europe, awoke one morning to find that more than three-quarters of the electric power generated for their nation that day came from renewable sources (the overall average is 24 percent, and rising). Germany decided a decade or so ago to get this done, to turn away from both fossil fuels and nuclear power. They have made enormous progress while I was tuned in to the naysayers here in the United States.

Government agencies own two-thirds of the NGS. The largest owner is the U.S. Bureau of Reclamation, part of the Department of the Interior. Ironically, the Department of the Interior is also responsible for the national park system and the Bureau of Indian Affairs. Another major owner is the Salt River Project, sometimes described as "quasi-governmental" and also as a "multipurpose federal reclamation project." Then there's the Los Angeles Department of Water and Power, which is negotiating to divest its share of the plant and use other, cleaner energy sources to comply with clean-air laws in its home state.

So whenever the federal government decides to get out of the dirty power business and conform to its own environmental regulations, Canyon views will clarify, regional air will be healthier, and—if our leaders plan for it—there can still be jobs. The Massachusetts report puts it this way: renewable energy "is only mining the sun, and harvesting the wind." That's a lovely sentiment but not, of course, the only basis for this prospective deal. The trade-offs will be costly. But then, the status quo at the Navajo Generating Station and our other fossil fuel burners costs far more.

CHAPTER EIGHT

Mount Trumbull

*They pretend we can have it all—grazing and healthy
ecosystems.*
    Thomas Fleischner, biologist

We're juddering south along a dirt road from Saint George, Utah, over
the Arizona border and onto an austere blanket of sagebrush and juniper desert. It unfurls steadily to the horizon, anchored against the rock-
ing wind by solitary broken buttes.

My cell-signal umbilical soon tapers to nothing. The very red vehicle I command bears reassurances, though: it's a Wrangler, badged as
Trail Rated. This model is a Rubicon too, as if equipped for some fate-
ful macho transit.

The mythmaking here is just a string of see-through marketing
banalities. Funny, though, how they can still drive consumer behavior.
Scenery of the kind we're rolling through is also loaded with myth. It
conditions, heavily, how we think about the people who use this landscape and govern it.

Well, the car's okay, but Rent-a-Jeep rubber is usually thin as gift
wrap, which matters on these roads. During a prior desert excursion
the gift was two flats in two days that nearly stranded us, overnight
or longer. This time: eight-ply, waist-high tires, two spares, and a
big cargo of drinking water. We will see three other cars and no
pavement in ten hours and 150 miles today. My cousin Howard and

I plan to backpack on this nearly deserted terrain, which has been described as one of the most remote landscapes in the United States that isn't Alaska.

This land is not really deserted. A caucus of charcoal-colored cows suddenly looms near the road; calves, and mamas with yellow ID cards stapled onto their ears. They are stolid munchers, moving through the forage with heads down, but shy enough of tourists with cameras to trot away as I approach.

A bit farther on is a wooden sign, cheerily ventilated by bullets and buckshot, that announces ownership. Though as it turns out, the ownership is contested.

BUREAU OF LAND MANAGEMENT
YOUR PUBLIC LANDS
USE—SHARE—APPRECIATE

The BLM, an agency scarcely known to most Americans, administers some 386,000 square miles of public land, most of it in western states, far more than any other federal agency. Above 60 percent of it is leased out for grazing. That is 230,000 square miles—something like the land area of Pennsylvania, Ohio, Virginia, Indiana, Maine, South Carolina, Massachusetts, and New Jersey combined. The Forest Service leases grazing rights on another 148,000 square miles in 29 states. There go New York, West Virginia and Washington, too (see Map 5). Even the National Park Service allows grazing at thirteen parks, including 1,200 square miles of the Glen Canyon National Recreation Area.

The scene I'm navigating now is a desiccated monotony to most eyes. Variations on it blur past us along the roads of much of the flatland West. A lonesome clutch of cows really seems to belong here. If we notice them at all, most of us feel okay about these animals for a lot of reasons. We might figure that if a hard-working family has found a way for cattle to eke out a living from that sparse gray-green foliage, then what an ingenious use for a hot, empty void!

A closer familiarity with the surroundings is in order, though. This particular patch of desert is an extension of the natural life of the Grand Canyon, now and even more emphatically in the future—the trees, the bugs, the wildlife, the dirt. Globe mallow, blackbrush, sagebrush, Brigham's tea, bighorn sheep, wolves, cats, condors and Gila monsters all are the natural tenants of what will soon be a rapidly, chaotically shifting habitat.

As for those innocent cattle, they are fairly recent arrivals, natives of Eurasia, and not a natural part of any North American ecosystem. They have become major players in the increasingly urgent drama of how, or whether, the United States will allow its public lands to adapt to the ongoing enormity of climate disruption. The wide vistas moving past my windshield are an invitation to a long-overdue national cow conversation.

"The western U.S. is looked upon as an arid place, but it has amazing ecosystems that are unique to the Rocky Mountain and Grand Canyon regions," the geomorphologist Robert Beschta told me. "Everybody knows the Pacific rainforest has incredible biological diversity, but it doesn't match what we see in these intermountain areas. Plant communities are the underpinnings of these ecosystems—the animals and everything else that lives out there," he said. "Now we overlay two major stresses. One is climate change, which is affecting large areas incrementally and will shuffle plants in ways that perhaps we can't even comprehend.

"On top of that is this other stressor: grazing, on huge areas. It has been going on for more than a century. So between the two, we're kind of heading over a cliff. Species are being stressed by climate to find new niches and new locations. At the same time, cattle are preventing them [from] growing and producing seed. It's a double whammy for plant species and the animal life that depends upon them to continue to exist. That's the scary part of where the current scenario is taking us."

A couple of hours in, we enter the Grand Canyon-Parashant National Monument—hereinafter just called the Parashant—which consists of 1,500 square miles of this isolated landscape. Grand Canyon National

Park is on its southern and eastern borders, Nevada to the west. It is eerily devoid of human residents, though not human impacts. This national monument is as big as the national park, potentially doubling its protected wildlife habitat.

From west to east the Parashant is escalated skyward by several seismic faults. They have created a sixty-mile giant's stair, a series of sheer rock faces alternating with broad flatlands that link the Mojave Desert to the Colorado Plateau: Grand Wash Cliffs to the Sanup Plateau, upper Grand Wash Cliffs to the Shivwits Plateau, Hurricane Cliffs to the Uinkaret Plateau, and an eroded volcanic hump called Mount Trumbull.

The temperature at the bottom of Grand Canyon, not far away, can push well past ninety degrees at this time of year—mid-April. But when we finally reach the trailhead at the base of Mount Trumbull, the wind has turned frigid. At eight thousand feet, its summit is the highest point on the Parashant. By midafternoon, it will snow here.

When climate change pushes parts of the intricate web of life out of southern Arizona and toward the Grand Canyon, and out of there and on to still higher or more northerly places, this landscape will be one of the way-station refuges. That will be possible only if it still resembles a natural ecosystem, however. Under siege from cows, invasive grasses, fires, and drought, the resemblance grows fainter.

Much will depend, then, on seeing the park and this monument and all public lands as integral, as broad pathways into the future. The Parashant could be a model for how we need to reassemble the puzzle. Just now, it is a cautionary example instead, in political as well as biological terms.

The monument is jointly administered by the Park Service and the BLM, and it includes four designated wilderness areas—Mount Trumbull is one. The Wilderness Act of 1964 allows Congress to grant these areas a high level of protection. Here's the lyrical dictum that is the heart of the act: "A wilderness, in contrast with those areas where man and his own works dominate the landscape, is hereby recognized as an area where the earth and its community of life are untrammeled by man, where man himself is a visitor who does not remain." But in a compromise judged necessary at the time for its approval by Congress, the act also grandfathered in widespread, albeit conditional, cattle grazing.

On the Mount Trumbull Wilderness Trail we climb through pinyons and junipers into a rare forest of never-logged "old growth" Ponderosa pines that the wetter, cooler climate up here favors. When the sun warms their pitchy pale-mauve bark, it smells like fine vanilla, but you have to sniff hard on this cold day.

Timber, copper, gold, and livestock forage had lured settlers and speculators here by the 1870s, but some of these trees are more than five hundred years old. They are so big, and this mountain so remote and rugged, that the sawyers had to let them stand. A steam-powered sawmill was erected on a lower and more accessible area of the slope, where most of the Ponderosa forest was taken down. The lumber was hauled away by mule wagons to build a handsome Mormon temple that still stands in Saint George, sixty-six miles north.

What remains at the mill site and on up the mountain—after a century of diligent quenching of natural fires—are dense thickets of spindly young stems that suck water away from the larger trees. They're often called "dog-hair" forest, and they change the dynamics of wildfire. In a natural forest creeping, smoldering fires clear out shrubs and debris from time to time, only scorching the lower trunks of big trees and then moving on, leaving most of them healthy.

But the current forest of skinny trees and shrubs and built-up debris feeds hotter, more destructive fires. It is also a ladder for flames to climb into the treetops. They become "crown fires" that accelerate even faster and destroy, rather than singe, the big old Ponderosas. The lengthening drought here has sparked high-intensity wildfires on Mount Trumbull. Many of the aged sentinels are now black snags, and the rest are at risk.

A flat, forested saddle connects the Mount Trumbull and Mount Logan wilderness areas. It's a research site, where scientists of various disciplines, mostly from the Ecological Restoration Institute at Northern Arizona University, have tried to divine what might help restore a natural, presettlement mix of vegetation. Their aim was to bring the Ponderosa ecosystem back to vigor and stability after a century of hard human use and mismanagement and make it more resistant to drought, fire, and climate disruption.

Restoration ecology—research on putting compromised or broken ecosystems back together—is inspirational not only in its aims but also its results. And these are the kinds of people—the 'ologists—to whom I habitually turn for answers about what's going on with natural habitat.

Theirs is a poignant goal on this national monument and in this sort-of wilderness. There are no longer any pristine areas within the whole vast Parashant. The BLM's district office told me there's no place where you or I could even find a single area that is rated in "excellent" natural condition. A principal basis for the "excellent" rating is this: that the diversity and abundance of plant species is at least at 75 percent of its condition before European settlement. But another baffler is that no one really knows what the vegetation was like back then, I was told, nor how heavily it was influenced by Native American culture.

A few years after the monument was established, one Park Service botanist set out to find "The Ungrazed Piece of Land"—a lost legacy of the presettlement ecosystem. It stayed lost. She even helicoptered up on top of a high butte, but herders had built a way for sheep to get up there, too. Later, two U.S. Geological Survey researchers continued the quest with an experimental model to guide them to likely spots to discover a mix of natural vegetation in its original condition. They failed to find even one. Ranchers were consulted for leads. None panned out.

Their quest touched on an issue for natural areas management that can easily become strident, if not infuriating, among planners. It is that humankind has worked broad changes on most of North America since long before Europeans arrived—and "wilderness" outside all human impacts is, especially with climate change, a concept but not a reality.

So if we're going to restore an ecosystem, to what era do we propose to turn back the clock? Shall we recreate a landscape that Native Americans burned on a regular basis, complete with the plants and animals that thrive under that regime? Do we jump back to the time, as the last Ice Age ended, before humans showed up at all? These are questions about absolutes. They have sometimes thrown recovery efforts into disarray, especially when they are used as political leverage to argue that, hell, since there's no achievable "natural" condition, then why should we do anything but keep on keeping on?

Uncertainties about landscape history are important, often inescapable. They can also be strikingly marginal to immediate needs. At the Parashant it's like arguing over the authenticity of candlesticks at the Chartres Cathedral when the roof is on fire.

The continuing damage to natural systems on the Parashant, and to every natural system across the United States, can be halted and often reversed, despite our lack of precise knowledge of where that will lead us. Research by David Tilman and others has found that ecosystems with a broader diversity of native species are more resistant to many kinds of environmental shocks and stresses, including drought and disease. It's not, perhaps, astonishing that intact and functioning natural systems are more stable than damaged, incomplete ones. But ecosystem health isn't either-or, any more than your own health. There is a spectrum of possibilities. Restoration isn't the pursuit of some crack-brain's Eden, but of resilience and continuity.

Roughly half of the Parashant is in the Mojave, the driest of the four types of deserts in the Southwest. In 2015 that part of the monument saw almost no rainfall—the average is only five inches. In healthy condition, though, even the Mojave is rich in plants, animals, and insects.

We have fleeting glimpses of what some of the higher desert here was like before the onset of grazing. In 1879 the Canaan Company established a dairy ranch at Oak Grove on the southern section, under the firm hand of a man named Albert Foremaster. A hundred years later his daughter Florence reminisced, "Daddy said that when they first moved to Parashant it was like a meadow everywhere, but overgrazing has changed all that." Indeed, the impacts of sheep and cattle on this dry, fragile landscape were so severe that many pioneering ranchers gave up and sold out only a few years after settlement.

The Mount Trumbull restoration ecologists have been at their research for twenty years—a couple of generations of scientists and their graduate students pursuing a suite of studies. The work uses controlled fire and cutting to thin out that dog-hair forest on selected sites. Then there's patient long-term monitoring to gauge the results. With fewer small trees more water is available for the larger ones, and the thinned forest is less susceptible of destructive levels of fire. There is evidence that more diverse natural vegetation does return.

It's important research that could help us figure out how to manage tens of millions of acres of other forests that are increasingly burning or dying off, from California to New Jersey, New Mexico, or British Columbia.

These scientists have to account for as many variables as possible that might skew their results: fire, heat, rain and snow, the soils, and the history of the landscape. Some factors are controllable, some not. The distinction matters here. Here the cows are a constant, a given, a factor conceded to be beyond any control.

This site between two wilderness areas ought to be a fine place for research on restoring natural processes. The two federal agencies—the Park Service and the BLM—that patrol the monument promise careful, science-based, ecosensitive stewardship for all of it. From the Park Service's mission statement: "The Monument is a model of research, investigative studies, and scientifically based management that guide the restoration of ecological processes.... Conservation and restoration of habitats that support sustainable populations of a full range of native species, including predators, are emphasized."

But despite those assurances, the research area and even the wildernesses on this monument are grazed by cattle under permits issued by the BLM, and most of the predators have long since gone missing. Four-fifths of the entire million-acre, publicly owned Parashant is chewed by privately owned herds—as many as 3,500 animals.

A big sign at the only mapped trail that leads into the Grand Wash Cliffs Wilderness warns that no mechanized equipment is permitted. A heavy chain is slung across the trailhead into this sanctuary of wildness. But the chain can be parted, and the ostensible hiking trail turns out to be a dirt road with a sixty-foot-wide easement. It is exempted from the no-machines rule if you run cattle here. On the day we backpacked in, we followed the tracks of a bulldozer that had been sent over the road to keep it wide open for ranchers.

On the list of serious cattle impacts on public lands, start with these obvious ones: they eat native vegetation such as grasses and seedling trees, they compete with wildlife for that forage, and they alter the natural mix of species by favoring some plants over others. They tend to gather near sources of water, especially in arid places. A long train of research documents that the presence of cattle there drives away other wildlife; strips away vegetation; denudes, erodes and collapses stream banks; pollutes the water with silt and dung; and diminishes or destroys fish populations.

These heavy animals also compact the soil and pulverize its delicate, nitrogen-fixing biological crusts, so topsoil succumbs to the sheet erosion of wind and rain. Biocrusts are essential for vegetation and wildlife on many kinds of arid landscapes. (Commonly said: cattle are just the modern equivalent of buffalo herds. This combines myth and rationalization. There were very few buffalo west of the Rockies, and they were widely scattered and occasional, rather than resident chompers.)

Cattle also, as a wealth of research affirms, spread cheatgrass. We met this native of Europe, Asia, and Africa in an earlier chapter. It can be found on most of the monument outside the Mojave (which is overrun with red brome, a different invasive). Cheatgrass seeds cling to bovine fur and hooves, but, mainly, cattle just like to eat this plant.

Its hundreds of seeds can pass through their complex cow plumbing —a process that scientists have also studied meticulously—and then out onto previously unaffected areas of land. The cattle's hooves conveniently press seeds into the soil and create pockets, where water gathers to help them germinate. "There's no question that any kind of livestock grazing transports cheatgrass and many other kinds of invasives, because that is the way they are designed to disperse," Mark Wimmer, the BLM's acting manager for the Parashant, told me. Banishing cattle will not get rid of cheatgrass, but it can't be eliminated with cattle present.

So, unfortunately, after years of toilsome ecological restoration research and considerable expense up on Mount Trumbull, cheatgrass exploded on the scene, to the chagrin of the scientists. The likely role of the Mount Trumbull cattle herd in the invasion was pointedly raised in the research literature.

Over just a few seasons cheatgrass can erase native perennials like sagebrush—once a keystone of a reassuringly huge western ecosystem and now rapidly dwindling. An annual, cheatgrass burns often, wiping out the native plant communities. Its root system can suck most of the water out of the soil, to a depth of twenty-eight inches. This invasive can "shift the invaded ecosystem into an alternate stable state that is likely irreversible." And once established, it has proven all but impossible to get rid of.

"We like cheatgrass," one rancher told me, however. "Cows eat cheatgrass, when it's green. So it's not that big of a deal, to us. Obviously it's an invasive and it's not native, but it's good feed for our cattle." What about the fires that cheatgrass brings? "We're not totally against fire either, because it cleans out the trees and everything," he said.

One of the Trumbull scientists wrote that "If cattle grazing is, as our data suggest, promoting cheatgrass on the restoration site, then we must caution land managers against cattle grazing, even at low intensities, on Ponderosa pine restoration projects in areas containing significant amounts of cheatgrass or other non-native plants."

Cattle still graze Mount Trumbull, though. The acceptance of grazing as a given, even in a wilderness area, can pervade even "restoration ecology" research in this setting. "It's been a learning experience for us," the Ecological Restoration Institute's research director told me, "but it's, you know, real life.... The ranching interests are all across the West, and that's something that we need to incorporate into the planning and management of what we're doing."

This may look like pragmatism akin to treating lung cancer while acquiescing—with a silent shrug—in the patient's heavy smoking habit. But if you are forced to choose between research with cows or no research at all, you might make the same choice. Scientists don't dictate terms on this wilderness. As for the "restoration" effort, it is compromised, to say the least.

The remarkable conclusion of the BLM's own personnel about how the land has fared on their watch is not reassuring. A recent annual report declares that the condition of the range and vegetation on the whole of the Parashant is only "good to poor," though the causes are unspecified. That picture scales up to the agency's damaging national self-portrait.

The true condition of federal grazing land is not easy to know. The BLM has stated that it has no reliable system for characterizing the health of its landscapes, let alone grazing impacts. I spoke with one member of a team of researchers for the U.S. Geological Survey who tried to pull the BLM's scattered land evaluations together. He was hopeful that things may improve, but we went down a list of the challenges. Were the records decentralized among dozens of offices around the country? Check. Were monitoring criteria inconsistently applied between states, districts, and perhaps individuals? Check. Is the number of cows on a grazing allotment self-reported by the rancher and usually unverified? Check. Are large areas unevaluated, in any case? Check.

The health of 38 percent of the grazed public land is not known at all, the agency has admitted—the samplings are incomplete or nonexistent or missing for other reasons. About a third of the 230,000 square miles of the allotments is listed as "all standards met," though aerial photography flatly contradicts that in some cases. On the other hand, BLM personnel told me, the sampling process can mislead in the other direction. Some land the map shows as not meeting health standards may actually meet them.

So the national picture is decidedly unclear. But the BLM's own sampling of grazing impacts is grim (see Map 6, which includes small units of state and private land). It indicates that more than seventy thousand square miles may be damaged by livestock so severely that even the agency's own fuzzy land health standards are not met. The condition of well over a third of this grazed, publicly owned vastness is simply, sadly unknown.

Here's a summary of a couple of dozen research findings on grazing impacts at a variety of sites in the western United States, from a peer-reviewed article by the Arizona biologist Thomas Fleischner in the science journal *Conservation Biology*.

Those research studies discovered that with grazing cattle around, the density of small mammals was cut by a third and species diversity by half; perennial grasses and some palatable shrubs decreased dramatically; annuals diminished by as much as 60 percent and perennials by as much as 29 percent; in a sagebrush desert ecosystem, the diversity of plant species was cut by two thirds; ten grass species in a mountain canyon were all but wiped out; plant cover along streams was cut by more than half; and fewer streamside migrant bird species survived. Livestock also trampled desert tortoises, competed for their forage, and damaged their burrows. Mojave tortoises are an endangered species whose numbers are plummeting.

When there's no grazing, on the other hand, desert grass increased by 110 percent after thirty years of protection; plant species richness increased after removal of livestock in semidesert grassland; species richness increased 250 percent nine years after grazing was halted; shrubs quintupled and willows octupled after grazing stopped; songbirds, raptors, and small mammals increased 350 percent after eight years of rest from grazing; ducks and all terrestrial nongame birds were more abundant in ungrazed sites; there were twice as many lizards, and they were 3.7 times bigger on ungrazed sites; and trout production increased 184 percent when grazing was reduced or eliminated.

Some research suggests that livestock grazing may be the major factor negatively affecting wildlife in eleven western states, which compounds the problems of adaptation of these ecosystems to climate change. The widespread and ongoing declines of many grassland bird populations damaged by grazing are "on track to become a prominent wildlife conservation crisis of the 21st century," one study concluded.

Assume, though, that this list is suspect. Perhaps all those studies were poorly done. Or the list is incomplete and stacked against the benefits that cattle may bring. There is, after all, a grazing-oriented school of thought, not well substantiated so far, that claims that it is good for natural systems: cows can spread native seeds or regulate cheatgrass by eating it, under certain conditions. Or, if their numbers are few and they are very carefully deployed, cattle might not have much of a negative impact on vegetation. Indeed, a couple of conservation groups are collaborating with ranchers here on the Colorado Plateau to see whether lighter grazing, seasonally rotated and carefully monitored, might work with an effort to reclaim soils and streams.

Many, probably most, ranchers are certainly convinced. "We are here to take care of the land, and we use cows to do it," one Parashant rancher told me. "It's been proven in many places that cows help freshen the ground, so it's better, because we chained a lot of the trees, opened the country up, allowed grass to grow."

Ranchers have also discovered, among the remarkable assertions on a rousing political website, what purports to be research showing that desert tortoises thrive on cow dung. This would not be the only record of a species that relishes BS but, tortoise research experts have assured me, there is no evidence for it.

Gary Sprouse, a Nevada rancher with leaseholds not far from the Parashant, has one of the largest assemblages of BLM and Forest Service grazing permits in the United States—nearly 1,200 square miles of public land—and decades of experience. "When you talk to these rangeland biologists, I'm sorry," he told me. "I have little respect for them. I've seen them all my life. Most of the time I don't agree with a thing they have to show. And I think most ranchers are the same way. If we'd leave it up to the ranchers, this country around here would look good," he said.

Adopting these alleged uncertainties for the sake of argument, the question yet remains: why do the risks rest entirely on the native plant and animal species that inhabit, or could inhabit, these public lands? The burden of proof might instead rest with cattle operators and their BLM provisioners, since many such habitats and their natural systems are already rare and declining rapidly. For now, even if you're a committed cow optimist, public lands are an enormous experiment. Here, our high-risk, high-loss grazing goes forward indefinitely, after more than a century, over tens of millions of acres public land on behalf of private interests. And there's abundant research evidence of where it has led.

This upside-down experiment could continue well if it were rerigged, though. Lingering questions about the impact of grazing could be resolved to everyone's satisfaction. Remove the cattle from our battered public lands and begin to restore them to health wherever possible. Let the ranchers graze their own private land any way they like and tout their results. If any of the hoped-for benefits of grazing transpire over time, well then, what a wonderful thing to discover.

"Shifting the burden of proof ... is warranted due to the extensive body of evidence on ecosystem impacts, and the added ecosystem stress caused by climate change," one study concludes. Grazing should be allowed only where careful research comparisons and monitoring have demonstrated that it is "compatible with maintaining or recovering key ecological functions and native species complexes."

That recommendation has not been heeded by federal land managers, though they have a legal obligation to discontinue grazing where it damages the land. "It's a classic, emperor's-new-clothes kind of thing," Fleischner told me. Even among some scientists, "Grazing 'comes with the terrain.'" But in a very literal sense, it did not come with the terrain. It was introduced, and has been disastrous for the terrain. "What is not valid is when we pretend that there are no impacts, no trade-offs, no ecological costs," he said. "Grazing proponents don't say grazing is the lesser of evils or the least damaging thing to do. They pretend we can have it all—grazing and healthy ecosystems."

Back on Mount Trumbull, another potential cause of the spread of cheatgrass should be noted. It is the occasional presence of the very few human visitors here. Boot soles may indeed carry seeds. If it prevents the spread of cheatgrass, human visitation should be tightly monitored and those boots cleaned, a practice already followed in other natural areas that are vulnerable to alien species.

But it is hard to imagine hunters, hikers, and scientists eating wads of cheatgrass and other invasives and then shitting seeds all over the Mount Trumbull, Mount Logan, Paiute, and Grand Wash Wildernesses and the rest of our grand national estate of public land—at least not with anything like the forceful dedication of the hundreds of thousands of cattle we now allow to roam the premises.

CHAPTER NINE

Cash Cows

From an economic point of view, you have a serious environmental problem that's encouraged simply by stupidity.
Thomas Power, economist

To cross the landscape of the Parashant National Monument is to move among vague layers of public and private supervision with varying rules of use, observed or maybe not. An intricate pageant of local economic interests and belief systems, of the government and its influencers—it all plays out over this broad landscape like fleeting, morphing cloud shadows. And all of it is nearly as irrelevant as the shadows, if we are to plan for the survival of the natural landscape. For ecologists the patterns of life here are not a jigsaw of jurisdictions and warring interests. They are one entity, indivisible as the sky.

Near a remote crossroads is an outpost of Parashant history, a one-room schoolhouse that is now a museum—the well-cared-for remnant of the former community of Bundyville. The last permanent resident of the area left in 1984, but we roll past the corrals and fences of the Lazy S-O, a working cattle ranch whose owners live in far-off Saint George. It has been held for five generations by a family whose patriarch is Orvel Bundy, a fairly common patronymic around here.

Ranchers are not their stereotypes. Some are devoted, even spectacular, conservationists. This branch of the Bundys nonetheless echoes the feelings of many, if not most, public-lands ranchers, and it's useful to know their view of the world. For now, it dictates much of the future of the Grand Canyon region and beyond, from here to Canada.

The Bundys are keenly aware of the century-old federal legislation—the Antiquities Act—that gives the president the power to designate federal lands that include "historic landmarks, historic and prehistoric structures, and other objects of historic or scientific interest" as national monuments. "Scientific interest" was the blank check President Theodore Roosevelt cashed to make Devils Tower, Wyoming, the first national monument.

In 1908 he used it again, circumnavigating a lethargic Congress and furious Arizona legislators to shield Grand Canyon from private interests, as a national monument. That was challenged by a mining company, but the president's prerogative was upheld by the U.S. Supreme Court, then and in a subsequent ruling. (Later Congress made Grand Canyon a national park.) Presidents George W. Bush and Barack Obama used this same executive power to create new national monuments, including Pacific Ocean marine reserves that are larger than all of the United States' national parks combined.

And back in the year 2000 President Bill Clinton created the Grand Canyon-Parashant National Monument. That move ignited outrage among some in Arizona and Utah. Discussions about the future of public lands in the West have always been so, but they became even more radioactive after this and other Clinton-era monument designations. Someone warned Orvel Bundy not to go near the press conference that proclaimed the monument—it featured Clinton and Interior Secretary Bruce Babbitt at the South Rim—because he might be shot by the government.

Shooting or being shot at have now become part of our civic discourse around western public lands. As the need to consolidate them becomes more urgent, a countermovement is underway that seeks to transfer federal control to the states or to privatize and atomize them altogether. Both its law-abiding proponents and its violent fringes are gaining strength and may well prevail.

About 1 percent of the Parashant is privately owned. The rest has been federal land since the Paiutes were banished, and nearly all of it has been grazed under federal permits for generations. The ranchers and their political allies, including seven out of eight congressional delegates and Arizona's governor, objected to the monument designation.

"I first encountered [this region] in the 1950s and was struck by its beauty and remoteness. I've been in love with this landscape for a long time," Babbitt, the interior secretary, told a Flagstaff public hearing when the monument was under consideration. He is a former governor and an Arizona native, from a family of longtime ranchers. "Before he became who he became, we thought he was our kind of people," Orvel Bundy told me.

As recounted in an admiring article in *Backpacker* magazine, environmentalists celebrated the possibility of a higher level of protection for the Parashant, but the auditorium at Flagstaff also heard angry ranchers who heckled Babbitt, shouting, "Why don't you just go back to Washington and leave us alone?"

Coverage in the magazine *Range* was of a different tenor: "Babbitt gloated and rocked on his heels like a pompous Mussolini agreeing to a timid few questions from the press ... With a jowly imperial smirk, the Secretary paused a beat before answering." *Range* quoted Arizona Cattle Growers president Jed Flake, who "erupted in frustration. 'This isn't about protecting land, it's about vying for political PR points with an uninformed suburban public. The land in question is already protected by the Bureau of Land Management, and is depended upon by hundreds of ranch family members who, as stewards, protect the range and its sustainability." There are only twenty-eight grazing allotments on the Parashant, so even with their considerably extended families, "hundreds" of family members was an unlikely speculation.

"They're hurting people, not protecting anything from anyone," Orvel Bundy told a reporter after the monument had been declared. "I guess I feel like a Phoenix homeowner who just found out their home was designated as parkland. We're extremely scared."

In this statement, he neatly annexed public land as his home, with himself as home owner. This is common among grazing permit holders, whose children, too, are often raised to think of public land as part of their patrimony. It is also a source of great confusion in public discussion, this oft-heard, sleight-of-mind argument that a "federal land grab" is being plotted.

"We went to a meeting there in Saint George about this new monument," Orvel's son Bill recalled. "That was their big issue: 'We want to protect it for future generations.' And I looked at the guy and I told him, 'No, that's not true. Because what you did is you just took it away from my kids. The lifestyle that I know, they won't be able to enjoy, because they won't be able to do that. So you're not protecting it for future generations, you're just controlling it for your own interest.'"

Looking over the Parashant all these years later, though, it may be hard to figure out all the toxic rhetoric about ranchers losing their livelihoods and about the federal "land grab" (of … federal land) at the behest of those suburban voters. Mining is still allowed on the claims scattered around the monument, though none are active for now, and no new mines can be initiated. The ranches and the cows still abide, in just about the same numbers as before.

Uncontrolled grazing was reined in by federal regulation in the 1930s, after the ruin of both public and private land throughout the West had become a national preoccupation, thanks to the Dust Bowl. Billowing clouds of dirt from overgrazed and overplanted western barrens turned the skies black, even two thousand miles east in Washington, DC. They were called by one U.S. senator "the most tragic, the most impressive lobbyists that have ever come to this capital."

The new federal mission as specified in the Taylor Grazing Act back then was "to stop injury to the public grazing lands by preventing overgrazing and soil deterioration." In practice, however, federal personnel deferred so completely to cattle interests that they even offered to leave the room during one rule-making session. Grazing became less chaotic, intensive, and damaging, but it continued where it had taken place before—even in arid regions. The outcome is that despite eight succeeding decades and periodic calls for large-scale interventions, even its own assessments show that much of the land administered by the BLM is, in ecological terms, a near-ruin.

An example from just one region: "Seventy-five million acres of public land are at stake and the clock is ticking.... A large part of the Great Basin lies on the brink of ecological collapse," the agency's acting director wrote in 2000. He cited invasive species as the cause. Curiously, eliminating or even cutting back grazing a principal means by which invasives spread—was not on the report's list of "goals and actions."

Many techniques have been tried to control cheatgrass and other invasives over the years—herbicides, imported insects, and mechanical removal, for example—but, so far, none holds out strong hope for restoring the native vegetation that continues to be lost. It may be fair to call these enormous areas "sacrifice zones," as some have. It is an interesting technical term that begs the question, "sacrificed for what, and for whom?"

For much of their history the BLM and the Forest Service have viewed their domains as livestock forage, timber tracts, hunting preserves, fossil fuel sources, or mining claims, with some public recreation thrown on. More recently, the mission to protect and if possible restore natural habitats and ecosystem functions has become part of the federal law that governs these agencies. Complex natural systems worthy of protection and perpetuation in their own right have more currency now, in word if not in deed.

In practice these agencies' mission—"multiple use"—as ordained by Congress and a succession of like-minded administrations of both parties, shows little recognition that some uses degrade or destroy others. Many scientists have observed that if multiple use is not to persist as multiple abuse, priorities will have to be imposed.

Babbitt, as Clinton's Interior Secretary, wanted to give the BLM more of a role in real, rather than lip-service, protection for natural habitat: "The nation's largest land-management agency ought to be induced to have a sense of pride," he said. He acknowledged that the BLM's culture was pro-industry and pro-cattle.

By the end of the Obama administration, much remained unchanged. One of a long list of examples: BLM officials in Utah were accused by their own employees of short-circuiting the rules for several years to rubber-stamp hundreds of oil and gas drilling permits in environmentally sensitive areas. But as the Trump administration gathered momentum, the Parashant joined a long list of national monuments whose protections and boundaries were in the cross-hairs of a hostile review. It was intended to "end these abuses," the president said.

Many of the Parashant ranchers—also called "permittees" by the agency—actually run fewer cattle than their BLM permits allow, often because of drought, at least according to the agency and the ranchers themselves. The numbers are typically self-reported by ranchers, and critics say the counts are sometimes lower than reality. Orvel Bundy, however, told me he was taught by his father to manage his herd size so that "there's always some grass left in the bank."

I asked his son Bill, how do you know when your allotment is overgrazed? "You can tell when the cattle are starting to fall off," he said. "Their physical flesh. All you do is look at the ground. If the grass is short, you're running short. It's a livelihood for us. If we eat ourselves out of house and home, then we haven't got anything. So we try to take care of it. A few years back we had a real drought, and we sold half the herd. We didn't have anything to feed them and we weren't going to eat the ground into nothing." This is common sense and clear testimony—that the herd sizes officially permitted by the BLM as safely within the "carrying capacity" of the land routinely exceed reality.

The agency is supposed to reach its own, independent judgment about carrying capacity and the restoration of rangeland health.

Notwithstanding its reams of condition assessments and management protocols, though, the BLM's real working criterion is often a simple one: let the ranchers decide whether the cows are too boney and the greenery too short. BLM personnel in charge of monitoring and oversight readily confirm that they usually depend on the ranchers to adjust their own herds as conditions change.

Not all of them are as careful as Bundy says he is. And no one's really there to monitor much, either, because another, different reality check is in order: even if it were feeling vigilant, BLM oversight can be slim to nearly nonexistent. Only one range technician is assigned to monitor grazing impacts on the more than 1,500 square miles of this monument. And environmental assessments that are supposed to underlie permit renewals are not a top priority at other BLM offices, where staffing is thin or oil and gas drilling applications have to be attended to. Those are offices, monument acting manager Mark Wimmer told me, where "you don't have the time to deal with grazing as much."

BLM personnel have laid out this seeming paradox for me: although they confirm that grazing permits are a privilege and not a right, they believe the agency really has little power to revoke them. "Managers may deem problems, but they are also given a mandate under the law to continue grazing," Wimmer told me. Once a grazing allotment has been set up, the land cannot be permanently retired from grazing. "To actually remove grazing from an allotment is not allowed," he said, and many BLM personnel would readily agree.

The BLM can't do it whimsically, it is true, but, according to many legal authorities, the relevant statutes, as well as a U.S. Supreme Court ruling, have explicitly reaffirmed that the agency has ample authority to reduce, revoke, or suspend grazing privileges for any of several reasons. Indeed, as one federal law says, public land may be withdrawn from grazing if it is found to be "more valuable or suitable for any other use."

Another of the many laws governing BLM lands states that their use should be decided considering "the relative scarcity of the values involved" and that the government should "weigh long-term benefits to the public against short-term benefits." And, finally, in managing the public lands the government "shall, by regulation or otherwise, take any action necessary to prevent unnecessary or undue degradation of the lands."

This means that on places like the Parashant, the foundation for reasonable change—halting cattle grazing, for example—has already been laid. The scarcity and value of healthy ecosystems here is not in doubt, and the BLM itself says the condition of the land is only "good to poor." That is true all across the BLM's huge country-within-our-country.

But it's well to ask, what would happen to the nation's supply of beef if public-lands grazing stopped? The answer is, almost nothing. Only about 1 percent of U.S. beef production comes from public lands. What about the income the federal treasury derives from grazing? Fiscal conservatives take note: as it happens, government administration of grazing costs the BLM about 6.5 times more money than the fees it brings in. In the most recently tallied year alone, taxpayers would have saved somewhere between $50 million and $80 million dollars if private cattle vanished entirely from public lands, by the BLM's own estimates. Other estimates for the losses from both Forest Service and BLM administration of grazing go as high as $125 million a year.

Would it create a lot of unemployment to halt public lands grazing? There are some fifteen thousand permittees altogether. For comparison, just three large casinos in Las Vegas employ more people than that. So suppose all public-lands ranchers were compensated for their grazing permit interests—perhaps using the $50 million to $80 million or more in annual federal administrative costs we'd save—and then "laid off." Far larger layoffs occur regularly in corporate America, and we hardly wince unless we're directly affected. In the single, fairly gentle year of 2012—the last time the U.S. Bureau of Labor Statistics compiled layoff data—employers initiated 6,500 "extended mass layoff events" that resulted in the separation of more than 1.25 million workers from their work.

The formula I've offered here may sound something like: let's dispossess thousands of ranch families, on the land for generations, and turn them all into casino workers. Is that the way we aspire to deal with our other minorities, other indigenous cultures and communities, in America? Substitute "Apache" or "Mennonite" for "Western cattle rancher" and measure the feeling.

But consider instead that all public-lands ranchers would not be laid off if their grazing permits ended. Some could still raise cattle profitably on their own private land, and others would be happy enough to get out of a hard life and an often marginal business if a path —some kind of compensation for their investment— were opened.

Thousands of others are absentees whose lives and main sources of income are far distant. They sleep in Los Angeles, not in the bunkhouse. Just as important is the fact that the people who control at least two-thirds of BLM grazing lands look nothing like the Bundys in terms of their financial wherewithal, and the amount of public land they lease.

Some grazing allotments are shared among several permittees; some permittees control several allotments; sometimes it's just one allotment, one permittee. That fuzzes up the picture available to us from BLM data, but it tells a hard-edged story, nonetheless. The distribution of this wealth of public land is skewed, to a staggering degree.

- About 15,000 individuals and corporate entities hold public-lands grazing permits.
- The top 1 percent of them—only some 152 permittees—hold interests in an astonishing expanse, about a third of the total BLM grazing land area: some seventy-seven thousand square miles.
- The next 5 percent of them hold about the next third of the total land area.
- The bottom 94 percent—about 14,000 permittees—account for only the final third of the BLM's public grazing lands.

The identities of the permittees are sometimes obscured by corporate names, but data provided by the BLM shows that the folks leasing great swaths of public land to graze their cattle and enjoy enormous government subsidies include the following:

Bruce McCaw's net worth was $925 million in 2016. His many ranches in Idaho and other states include at least 540 square miles of public grazing land.

Barrick Gold, a Canadian mining corporation whose market capitalization was at $18.8 billion in 2016, is also a U.S. public-lands cowpoke—with at least 695 square miles of grazing land.

Stanley Kroenke is a real estate magnate, sports-team owner, and one of the nation's largest private landowners, with a net worth of $8 billion in 2017. He and his wife, Ann Walton Kroenke—of Walmart's Walton family, with a $6 billion net worth that year—graze at least 460 square miles of public land in Montana and Wyoming.

The heirs of the late hotelier W. Barron Hilton, whose net worth was reported at $800 million in 2012, inherited public-land grazing permits in California and Nevada totaling at least 350 square miles.

Mary Hewlett Jaffe, an heir of the Hewlett-Packard fortune and member of one of the United States' billion-dollar families, controls public-lands grazing permits in Idaho of at least 250 square miles.

The J. R. Simplot Corporation is one of the largest U.S. public-lands ranching outfits. Its lands make up at least 2,800 square miles of grazing allotments in California, Idaho, Nevada, Oregon, and Utah.

The Southern Nevada Water Authority holds grazing permits for at least 1,450 square miles of Nevada's federal land in the region, and cattle are grazed "to preserve the ranching lifestyle," I was told by an Authority spokesperson.

The Forest Service allows grazing on about 117,000 square miles of our national forests. If anything, its assessments of the impacts of cattle on the health of those forests are even more vague than those of the BLM. There are 3,772 permittees for these grazing leases, and the pattern of ownership is strikingly similar to the BLM's: control of most of that enormous landscape is concentrated in very few hands. The top 200 permittees—just over 5 percent of the total—control about a third of the total Forest Service grazing land.

So maybe we need a cowboy litmus test to winnow the real ones from those who are merely absentee cattle barons living on our nickel. Then we can dispense federal subsidies, or buyouts of grazing permits, accordingly. What is the extent of your public-land leases and your net worth, we might ask. Do you live principally on your "home ranch" or anywhere near it—or at Tahoe or Biarritz? If we're deciding to sustain cowboy culture with public money on public land, let's make sure we're not helping out some other, less needful, kind of culture by mistake.

Okay, but then what would become of the communities whose local economies depend, in part, on ranching? Thomas Power is former chair of the Economics Department at the University of Montana and specializes in natural resources and regional economic development issues. After lengthy research, he concluded that livestock grazing on federal land is generally unimportant to local economies and even less so to state and regional economies.

Power's study analyzed all eleven western states with substantial public-lands grazing and a select group of counties within them. Grazing on federal lands contributes "only a tiny sliver" of income and employment in these places, he wrote—rarely more than 1 percent. Strong economic growth in the West, he concluded, will continue to depend on quality of life and "protecting the environmental integrity of public lands." Public-lands grazing is "clearly quite peripheral."

"These lands aren't very good for producing forage.... The whole thing's going on only because it's being given away," Power told me. "I think what's outrageous is that cattle are turned loose with minimal management. That becomes the dominant use of public lands, with serious but gratuitous environmental damage that includes shooting predators of cattle or even critters that compete for the forage. We label it 'multiple use,' but it doesn't turn out to be multiple use. Its dominant use is by the cattle, and from an economic point of view, you have a serious environmental problem that's encouraged simply by stupidity. If the forage were sold at a market price, a lot of those lands would not be grazed."

Only a few counties in the West depend on grazing even modestly, Powers said. In those places the impact on their economies—ranch-based jobs, income, or taxes—is small. "There can be no significant economic trauma associated with restricting or halting public-lands grazing," he said.

Orvel Bundy's operation owns about six square miles of land outright, but its 130 cattle graze thirty-four square miles of public land, part of it near the edge of Grand Canyon. Without the grazing permits, the ranch would be too small to be economically viable, his son Bill told me.

Each cow-and-calf pair is considered an "animal unit," and the Bundys pay the going BLM rate for all ranchers: $1.69 per animal unit per month for the grazing. With inflation factored in, that's nearly a dollar cheaper than it was twenty-five years ago. In fact, the current grazing rate is less than a fifth of what ranchers would pay for grazing on private land in Arizona or Nevada, according to the BLM. It may be less than what you pay to feed your goldfish.

Just now, at the livestock auction in Salina, Utah, a healthy cow and calf together sell for around $2,400. These are not necessarily, however, "cash cows" in the figurative sense, at least not for the smaller ranches. A public-lands ranch family's income from jobs in town frequently props up a fiscally shaky cattle operation. Ranching is often a lifestyle choice, not a big moneymaker. Orvel Bundy is retired. Bill Bundy drove a delivery truck before he retired. His two brothers work in carpentry and insurance, and the ranch also serves as a base for one brother, who is a hunting guide.

None of this squarely addresses the personal situation of an unknown number of ranchers who are truly tied to the land—economically, historically, and emotionally. Proximity and long attachment shouldn't be exaggerated, nor belittled. But some perspective is in order, the biologist Tom Fleischner said. I posed the question to him, what should we do about ranchers who would be displaced under a no-cattle scenario?

"I don't mean to make light of that," he responded. "There are some ranchers for whom that's a real thing. But I actually don't think that somebody like me should be expected to have an answer for that. Continuing the status quo of trying to graze cows in a place that can't support them is not an answer either. It is only going to get worse because of climate change. The sooner people try to come to terms with that, rather than expecting federal land-management agencies to try to accommodate these folks, the better."

Fleischner said that the question itself marks "how deeply ingrained the mythology of the cowboy culture is: when somebody points out that the landscape can't handle all these cows that are subsidized on federal lands, it is seen as a responsibility of that person to come up with an economic solution. But when a factory pulls out of a city, it's routine. The owner doesn't have to figure it out.

"Public-lands grazing is a shared problem of the nation, not a problem of a particular very, very small group of people. Historically, they have been coddled, basically because they are politically well connected." And perhaps for some other intriguing reasons that we'll take up next.

CHAPTER TEN

Sacred Cowboys

As an intelligent species, should we be doing that, "just because my grandpa did"?

Kathleen Harcksen, former BLM assistant manager

Debra Donahue lives with her dog, her husband, and a burgeoning crop of cheatgrass on three acres near Laramie, Wyoming. "I can see a BLM range allotment from my front door," she has written. Cheatgrass "carpets the steep slopes above the river, turning a characteristic shade of red-purple as it cures. Later in the summer, I watch the obliteration of a small perennial stream and the degradation of its steeply sloped watershed by the BLM permittees. At the same time I take twice-daily walks on nearby state and private lands from which livestock are excluded, and marvel at the diversity of grasses, forbs, shrubs, and lichens that were there. Cheatgrass, unfortunately, is present too, but the native vegetation seems to be holding its own against this ubiquitous invader."

Donahue has intimate knowledge not only of cheatgrass but also of the tenacity of folklore, tradition, and politics on the range, especially when they contradict federal law. A public-lands law expert, her book on grazing was written partly to lay some of the legal confusion to rest. But its reminder that public lands are legally under public control discomposed the cattle establishment. The irate president of the Wyoming Senate, for example, drafted a bill to abolish the whole state university law school where Donahue teaches.

Little has changed. A couple of years ago the Wyoming legislature again took offense on behalf of ranchers upset about anyone who might want to gather data about grazing impacts on the streams and soils of public land. They passed a law that declares it a crime to enter "open land," meaning public land, to collect data without the permission of the grazing permittee. It specifies fines and jail time. The law could arguably apply even to national parks. It requires that "illegally" collected water-pollution data be erased, unless it is used as evidence to help convict the trespasser. Headline on an op-ed: "Wyoming Doesn't Want You to Know How Much Cow Poop Is in Its Water."

Some observers have wondered about the source of public-lands ranchers' political clout, which extends very far beyond Wyoming or Arizona. There aren't many of these ranchers. Their influence isn't because of easily visible economic heft, even at the local level.

Picturesque, smaller-scale public-lands ranchers always form up in the media foreground when there's a fight over, say, new regulations to revive wrecked, unfenced streams in Oregon. But, as we've seen, most grazing lands are not leased by those grizzled, sunstruck landsmen in cowboy gear. Less picturesque and less engaged, these big absentee owners live elsewhere. They are corporations, lawyers, financiers, and other highly affluent folk who see a deal in a government handout, or perhaps like the notion of hobby ranching.

My favorite rootin' tootin' ranchers, though, are David and Charles Koch. The brothers hold state and federal public-lands grazing permits totaling about three hundred square miles, as part of their Matador ranches in Montana. The federal portion alone ranks them in the top 5 percent of all BLM grazing permittees. Their operation's mission is "preserving the cowboy way for generations to come," according to the corporate website. Their taxpayer-subsidized federal grazing leases, which are no doubt a fine investment, don't seem to square well with the Koch political portfolio, though.

We'll visit with the Kochs and their campaigns on public-land policy at greater length, further on. For now, and by way of introduction, only consider that they are both on the list of the top ten wealthiest people on the planet—worth about $35 billion apiece. And they have spent tens of millions of their fortune, over decades, ginning up political support for the idea that the dependence of U.S. citizens on government help is a really bad thing. As Charles Koch once said, "Repeatedly asking for government help undermines the foundations of society by destroying initiative and responsibility. It is also a fatal blow to efficiency and corrupts the political process."

Another enduring source of political voltage for ranching interests is cowboy mythos. As the economist Thomas Power says, "I've been really interested in why nobody seems to care about the economic arguments, because most people act as if they are interested in economics and economic rationality and all that sort of thing. But when it comes down to it, what people are really concerned about is protecting what they perceive to be an important cultural heritage: that people working the land for a living are independent, physically competent, engaged in economic activities that were originally associated with the European settlement of the western United States.

"There is real power in cultural symbols and a way of life that is supported by these federal programs," he told me. They're tied to an economy that doesn't really exist anymore. So it's the ghosts of the past insisting that we act as if our economy, in the early twenty-first century, is what it was in the late nineteenth century. It's crazy."

Most of us buy into it effortlessly, though. After all, we eat a lot of beef. Most of us grew up with video cowboys, those well-merchandised American archetypes. We see them with their six-shooters and craggy individualism when we gaze into the mirror of our national-history-lite: country-western music, checked shirts with snap buttons, all those big wide hats. Donahue's husband, a former cowhand himself, once scanned an overfond *New York Times Magazine* article about a rodeoing rancher family. "When New Yorkers see a cowboy, half their brains fall out of their ass," he explained.

Cows on federal land are only occasionally visible to city folk like me. The huge subsidies we invest to keep them there are not visible at all. "It's ludicrous," Donahue says. In a law journal article, she explains the real range of subsidies that grazing enjoys and that it exemplifies "the ability of narrow interest groups to 'capture,' or co-opt, the agency and institutions charged with their regulation." Along with all that cowboy mythology, she writes, this phenomenon called "regulatory capture" helps explain "why the unsound and anachronistic policy of federal public-land grazing has persisted into the twenty-first century."

The BLM does not put a sale price on its grazing permits because the public's land is not for sale, strictly speaking. But the permit has a market price that reflects the value of the public subsidy, nonetheless. Banks will loan money against it. When private ranches change hands, the seller's public-lands grazing permit does not revert to the BLM. Instead, it registers as part of the value—and the sale price—of the land, just as a taxi permit medallion in New York City makes the taxi a lot more valuable than the going price for a yellow car. The BLM awards the permit to the new landowner, almost automatically.

"And the longer this goes on," Donahue says, "the harder it is to convince people that public-lands grazing is not a property right. The law is clear that the BLM and the Forest Service can decide to assign the lands to other public use instead of grazing it, which is very much a private use. But that doesn't mean that legal authority is being used much."

Minuscule grazing fees, and the huge subsidy they represent, are not the biggest problem. "It's the damage the cattle do, and that damage is totally uncompensated," Donahue says. "Often it cannot be compensated, because it's irreversible." Sagebrush ecosystems are one example. They are among the most endangered in the West, partly because they have been devastated by 140 years of livestock grazing. "And in the West, I don't see any way to extricate climate change from the livestock grazing issue," she told me.

Yet another subsidy is provided by an agency, Wildlife Services, within the U.S. Department of Agriculture. It kills birds around airports for safety and kills harmful invasive species like wild pigs, among other tasks. The Wildlife Services Program also traps, shoots, and poisons predators and other kinds of animals at the behest of ranchers and other producers. "We pay to shoot them, sometimes from helicopters," Donahue said. "But we do it because there is this very vocal group, ranchers, that demands it. Helicopters aren't cheap."

"I think that if you want to interpret that as a subsidy, you could," Steve Kendrot, a Wildlife Services official, told me. "However, wildlife in North America is managed by the government for the good of the people. That includes all of the people, not just the people who want to see lots and lots of wolves or coyotes. When wildlife as a public resource impacts a private individual, I don't think it's a real big stretch to see the government as having a part in helping to resolve those problems."

Agreed. But it is indeed a big stretch to see mass killings of wildlife as the preferred form of resolution in so many instances, and on both private and public land. In just the most recent year for which data are available, the program killed the following, primarily on behalf of agriculturists, including ranchers: 299 badgers, a third of them unintentionally; 570 black bears; 796 bobcats; 61,702 coyotes; 2,651 red and gray foxes; 813 hares; 691 great blue, green, black-crowned, yellow crowned, tricolor, and little blue herons; 301 mountain lions; 2,557 woodchucks; 1,914 muskrats; 454 river otters, 390 of them unintentionally; 15,698 black-tailed prairie dogs; and 321 wolves (these are selections from a much longer list).

Not long ago, evincing concern about climate change and other environmental threats, the BLM convened a group of federal scientists to develop a national "Rapid Ecoregional Assessments" technique, "to more fully understand ecological conditions and trends; natural and human influences; and opportunities for resource conservation, restoration, and development." The project was funded at $40 million. The participating scientists were told by BLM officials to consider the impacts of "change agents" and evaluate them.

Then they were instructed that grazing was not going to be included among the "change agents" and was not to be part of the study. If that hints at light farce, the agency's defense when challenged enlarges the mood. After decades of assurances that it has been monitoring its lands to keep them healthy, carefully gathering data and stewardshipping, the BLM now demurred modestly. Its data on grazing and land health conditions are neither complete enough nor reliable enough to be included in its "ecoregional assessments," it decided. No curiosity was evident, either, about the substantial pile of research available from other sources about grazing as a "change agent."

The minutes of planning sessions for the assessment program show protests from some of the scientists present. "We will be laughed out of the room if we don't use grazing," U.S. Geological Survey ecologist Tom Edwards—a peer reviewer for the assessment— told the group. "If you have the other range of disturbances, you have to include grazing."

That was prescient. The ensuing uproar included a formal challenge to the BLM's scientific integrity by a public employees' environmental group. The agency investigated itself and found nothing amiss. Just the same, after producing six ecoregional assessments that put the hush on grazing impacts, it reversed course and included this "change agent" after all, in the seventh region.

All this is further evidence of a system designed to perpetuate the status quo—to accommodate the interests of the permittees and their political backers. Ironically, though, its track record on grazing buys the BLM little in the way of rancher love. Ranchers routinely fault the agency instead for being staffed with college-trained know-it-alls, tools of the nannying federal presence.

Gary Sprouse is a Nevada rancher who controls one of the largest collections of public-lands grazing allotments on the continent. He gave me one of the kinder assessments: "If we could just do away with the Bureau of Land Management and let the ranchers manage this land, it would be far more productive than it is now and it would look much better. Not that these people aren't good people—a lot of them are very nice and good to work with. But they're scared to death to make a decision. They're all managed out of Washington. . . . I don't mean to harm a lot of the people at the BLM, but it is the sorriest outfit. I mean, it's one of the bureaucracies that's ruining our nation!"

There are harsher symptoms of this disaffection. Even on the relatively peaceable Parashant, ranchers have sabotaged fences around protected areas and threatened to close public roads to protest BLM policies. Throughout the West the agency's employees were the targets of at least two hundred violent attacks during the most recently tallied decade. Former BLM director Patrick Shea told a reporter that he had always been apprehensive because of threats of violence when he traveled in the West. "Whenever I went to southern Utah and some other states, I had security people with me—it wasn't because of me as an individual but because of the office I headed," he said. The situation now is far worse, he added, because of vigilante militia networks and small-town leaders looking for votes.

Agitation against the federal presence often seems reflexive among much of Utah officialdom—an ideology of sour, rumbling paranoia that is constantly beamed at its citizenry. Washington County at one point created an official website inviting citizens to register opposition to BLM proposals for public land and helpfully included a form letter for them to submit:

"Dear BLM St. George Field Office, I write to oppose . . . " and a laundry list of "Areas of Critical Environmental Concern" follows: "I write to oppose . . . restrictions on all-terrain vehicles"; a multispecies management area that "could potentially limit mining, hunting, grazing, and OHV [off-highway vehicle] use"; and the introduction of the endangered California condor into the area "or any other protected species."

I discussed grazing issues with a BLM rangeland management specialist and a public affairs officer at the agency's Saint George office. They

acknowledged staffing and budget shortages and lamented that litigation over grazing "is a distraction of time and energy. It takes us away
from what we should really be doing." Rangeland health monitoring, I
was told, is complex and its criteria often change. No planning for climate change on the monument is underway: "We rely on the ranchers
to adjust herds for that." The rangeland specialist Whit Bunting also
reminded me, "We're managing for more than just livestock foraging
conditions." But the priorities for this unloved, understaffed, financially
strapped, and culturally cattle-inclined "multi-use" agency are also
evident.

The Mojave Desert tortoise provides an instructive example. The
species first arose about six million years ago, and when scientists began
to pay it much attention, they noted that its range was already severely
hemmed in by human activity. By 1990 it was declared a threatened
species, as populations fell rapidly. The latest and most detailed monitoring by the Fish and Wildlife Service found that in the *most protected
areas* of its habitat, tortoise populations are in free fall, down 32 percent
in just ten years. At that rate, extinction in nature beckons. Maybe they
could hold out in zoos, for a time.

Tortoise habitat is a war zone. Paved- and off-road vehicle traffic,
lethal vandalism—especially with guns—real estate development, and
large-scale solar-power projects all play a role. The BLM began spo-
radic research on grazing and tortoise mortality as long as forty years
ago that raised disturbing questions. The Fish and Wildlife Service has
long since called cattle grazing a "major threat" to tortoise recovery
and concludes flatly that these two land uses are incompatible.

On the driest, western edge of the Parashant is a grazing allotment
designated on BLM maps as "Critical Desert Tortoise Habitat." But on
that national monument land—advertised by the agency as a model for
the "restoration of ecological processes"—cattle still graze. And they
graze by permission of the BLM, despite requests from the Fish and
Wildlife Service to yield this "critical habitat" to the tortoise instead of
the cattle.

Kathleen Harcksen has also seen management of monument range-land at first hand. She is recently retired from a long career as a resource management specialist that began at the Forest Service and ended after a decade on the Parashant with the BLM, first as an assistant manager and then as a project manager. "They just put cows out there," she said. "There was never any look at doing anything different, at doing it right, or better, or to be in compliance with the law. BLM people work where they grew up. Their families often hold grazing permits. There's no out-side learning." Federal law prohibits BLM employees themselves from holding grazing permits. I asked the district office in Saint George several times whether relatives of BLM employees own grazing permits. That is not illegal, but certainly introduces a regulatory conflict of inter-est. District officials did not reply.

Harcksen characterized the agency's elaborate system of range man-agement assessments as a facade. "It's very detailed, that's right," she said. "They waste a lot of money collecting data that they refuse to look at, or they change the data, cook the books, and lie about it. These BLM managers are proud to say, 'Oh yeah, we work so well with the permit-tees! We have such a good working relationship with them! If they tell us to, we roll over and pee on ourselves, every single time we have to.'" The agency declined to respond.

The legal scholars John Leshy and Molly Mcusic summarize the track record of the BLM and the Forest Service in similar, if less vivid, terms:

> More than a century of regulatory history ... reveals how deeply sympathy for ranchers is embedded in agency culture—a classic case of regulatory agency "capture." Local ranchers press agency personnel, who respond predictably.
>
> Even if conditions on the public lands are seen as intolerable, federal land managers tend to tighten regulation only after years of monitoring, and in sedulous consultation with the rancher. They know that, if they show unaccustomed vigor, ranchers can call on Members of Congress and other political actors to intervene.

And even if the federal land managers limit the amount of grazing on particular tracts of federal land, history shows that they are extremely unlikely to retire tracts of federal lands permanently from livestock grazing altogether. Yet because vegetation is slow to respond and soils slow to rebuild in the more arid parts of the West, nothing short of permanent removal may allow that land to begin to regain health.

Harcksen seconds the motion: "There shouldn't be any grazing in any desert, anywhere. I don't know if these deserts can ever recover, at this point."

Harcksen, Donahue, Fleischner, and a long list of other experts have a variety of alternative visions for what public lands could become if grazing were curtailed and natural habitats became the priority, rather than remaining the beggar at the "multiple use" banquet. For a start, although the law as it stands allows the revocation of grazing permits for valid reasons and without compensation, there is precedent for buyouts, which go a long way to ease the process.

"As a matter of policy and equity, I've come around to the view that it probably makes more sense to buy out ranching permits," Donahue said. "Some ranchers have invested a lot of money in getting the grazing permit in the first place, by buying the home ranch. A lot of others have not. But they have that investment in the property and in their operation. If you cut the permit, you would basically deprive them of that source of income. So as a fairness, it just seems like they should be bought out."

That's an apt and generous proposal, but I would include a means test: "Any public-lands rancher with less than (pick your annual income) or (pick your total net worth) may apply for compensation for the retirement of their permit to graze public land." It would be paradoxical, somehow, to compensate the Koch Brothers and the rest of our list of millionaires and billionaires for the removal of their cattle from leased land that belongs to all American citizens. When I received allotment data from the BLM, it came with this legally impeccable notice: "Term authorizations convey no right, title or interest to any lands or resources held by the United States. Grazing use is a privilege, not a right."

Most of the biological diversity and complexity in the Grand Canyon region and throughout the arid West is focused near water. Dragonflies, aquatic beetles, caddisflies, springsnails, and other aquatic invertebrates gather around even the smallest springs and streams, along with a run of animals like skinks, voles, and birds. Tiny Oak Spring, on the southern Parashant, is only the size of a silver dollar and a half-inch deep. It drew fifteen species of birds in one survey—Scott's orioles, plumbeous vireos, several kinds of sparrows, and spotted towhees. They sit in the junipers nearby and then fly in, drink the spring dry and fly away until the little patch of water rises again.

Cows like to be where the water is, too, especially in these hot, arid climates. Grazing heavily impacts springs and watercourses. If, as on the Parashant at times, they are fenced to protect them from cattle, the fences may also exclude wildlife. To build and maintain fencing is an expensive proposition in any case.

Throughout the West, water is scarce, and it is the vital element most at risk from climate disruption. Water on grazed public land is routinely reengineered and repurposed by human intervention. The open water that flows on the Parashant, for example, is diverted to "tanks," or ponds, to benefit cattle and game animals for hunting. The springs are replumbed with plastic pipes and aprons that funnel the water, sometimes over miles, into covered catchments. Rainfall, too, is caught and stored in tanks called "guzzlers" that dot the landscape.

That's customary management, given current priorities. But the result is that many of these critical habitats and the species they support are wiped out. This is part of a process known across the West as "dewatering," and one in which cattle, the beneficiaries of the plumbing along with game animals, are a major factor. A cow eats roughly thirty pounds of vegetation and drinks maybe thirty gallons of water a day—more when it is enduring desert heat or lactating.

As a BLM project manager, Kathleen Harcksen worked for years on an inspiring example of restoration, along with a consortium of other agencies, nonprofit groups, and scientists. They rehabilitated a major source of flowing water—the artesian spring complex called Pakoon—near the edge of the Mojave.

Larry Stevens, an evolutionary ecologist and curator of ecology and conservation at the Museum of Northern Arizona, was also a principal on the project. "The Pakoon story is dramatic," he told me. "That was a devastated ranch. It had been used for alfalfa and livestock production for a hundred years, and then the manager turned it into an ostrich ranch. The entire landscape was just hammered by human activities." Nonnative bullfrogs, mosquitofish, and palm trees were thriving, along with an alligator named Clem.

The wreckage of house trailers and abandoned structures, a couple of cranes, four tractors, the carcasses of twenty-seven cars, and a rusting jumble of car parts were strewn over the landscape, along with hundreds, perhaps thousands, of miles of barbed wire. During the project tractor trailers hauled garbage away in a process that took several years.

"And we transitioned it from that complete mess," Stevens said. "We got the alligator out [and], brought in big equipment to recontour the landscape, and it erupted in life —a really fabulous response. The native vegetation reestablished and wildlife began to come in to use it.

"We worked to keep feral burros and cattle out of it, and that was met with quite a bit of resistance from the local ranchers," he added. One rancher sabotaged fencing around the project. Then, "Cows ate everything and crapped everywhere," Harcksen said. It took two months to remove the cattle, but the vegetation recovered.

They were also able to recharge the long-dead watercourse that is fed by the spring. "It had been dewatered for more than a century, and we recreated the longest-flowing stream on the monument," Stevens said. Every source of water on the old ranch had been excavated over the years and captured in ponds for irrigation and domestic use. The restoration project filled in the ponds and let the water seek its own course through the landscape again. The change brought in wetland species like bulrush, spike rush, and willows, as well as cottonwoods that grew to thirty feet in just two years. "They erupted into life," Stevens said. "It was just—it was just amazing." Pakoon is now, once again, an oasis in the harsh northern Arizona desert.

"Take out the fences, tanks, and pipes and let the riparian vegetation flourish," Harcksen counsels—an anthem that could apply throughout public lands in the West. Most of all: withdraw the grazing permits and remove the cattle.

"It's a national monument; it's a national treasure," Harcksen said of the Parashant. "It's a little tiny piece of land that doesn't need to be grazed. It should be made available for the vegetative and wildlife communities that can sustain it and the people who want to recreate there. Bighorn sheep, quail, turkeys, [and] predators like wolves and mountain lions could all prosper. With more prey you would have more predators. You would have a healthier system."

But, she mused, "we are predators too. And we are part of the system." It is apparent, in fact, that we humans have become the uberwolves, and in so spectacularly unbalanced a way that in some places even the vultures are vulnerable to extinction.

"So, what role do we play?" Harcksen asks. "How much of our intelligence do we use? How much of that public land should we be using to produce that 1 percent of beef? Right now we're using all of it. Even in systems like the desert that are brittle and that maybe we have broken. As an intelligent species, should we be doing that, 'just because my grandpa did?'"

Many public-lands ranchers, of course, have a hardened resolve to continue to do it. Orvel Bundy's great-grandchildren are frequent visitors to the work of his Lazy S-O ranch and its federal grazing allotments, which they regard as part of their family's heritage. They rebuilt the ranch house a few years ago, though no one lives there full-time. One wall is dressed out with planks from the old home place. When I stopped by, Orvel was just coming in with friends from a visit to the cemetery down the road. The family has run cattle here since 1916. Orvel Bundy says he was always told that the grass came up to a rider's stirrups back then, "a sea of waving grass," though ranchers tend to attribute the loss of that richness to a drying climate rather than to grazing.

The family does not think particularly well of the BLM and its meddlesome ways and certainly not as much of an ally. "There's always rumor that they're going to take this away from us. So we're always on edge because we never know who might get in as president or who might do something that jeopardizes our way of life or our living," Orvel's son Bill said. On the evidence thus far they have little to worry about, but the Bundys have prophesied since the creation of the Grand Canyon-Parashant National Monument that sometime soon the government will try to make it a national park.

That may be an agreeable plan for many, this author included. But ranchers are resilient, highly effective combatants. They fight on their home turf, with astronomically rich allies. "The place has been in the family quite a long time now, and we're not going to give up on the first drop of a hat," Orvel put in.

The national park model for the protection of natural areas, in which permanent human inhabitants are removed from the land, is sometimes referred to by critics as "fortress conservation." As "America's best idea," though, its successes here have been emulated around the world. Where the removals have been contested, on the other hand, as at Shenandoah National Park in Virginia, bitterness can result that persists for generations.

The original boundaries proposed for Shenandoah when it was created in the 1930s were far more expansive than the park is today. They were supposed to be achieved through purchases and land swaps over time. Instead, a larger park was fought to a standstill by hostile communities on its periphery. The park now is a beautiful but biologically starved place in some ways, a skinny and constricted hundred-mile-long ridge. Residential development is accumulating on its edges, foreclosing forever any possible expansion.

It's a fainter sentiment among urban dwellers, especially those of us who move fairly often. But in settled agrarian communities, identity is often rooted in the landscape. Its grip is forceful, enduring, and occasionally protective of the same values that environmentalists champion. When something threatens the connection between land and longtime resident landholders, the response may go deeper than mere economic self-interest or self-serving ideology.

Patricia Wilson Clothier's family wrested a marginal and physically hazardous living out of the towering rocks, dry arroyos, and cacti of Texas's remote Big Bend region during the Great Depression, for example. When talk of a new national park began in the late 1930s, some were happy enough that a willing buyer might finance their exit from very tough country. But "those who lived on and owned their land came to love it as home," she wrote. "We cared for the desert, the mountains, and this special way of life. We felt violated and did not want to give up and relocate.... These times for us seemed like the approach of a storm at night." Her memoir carries anguish from the loss, more than a half century later.

It's possible to honor these ties to the land and still conclude that cow culture is transforming enormous reaches of public land into an irrecoverable barrens and has already crowded many wildlife species to the edge of oblivion.

An alternative to the "fortress conservation" model, implemented in some places in Africa and South America, is to incorporate indigenous tribal folk into the natural life of the protected areas they inhabit. This offers the tribe a livelihood and a stake in the health of wildlife and their habitat.

The Masai people of Kenya and the Ovahimba of Namibia, for example, are emotionally involved with their cattle too, though the tenor of the relationship may be different than for U.S. ranchers. In one scene in the revelational documentary *Milking the Rhino,* a herder croons reverently, extolling the beauty of a beloved cow. The film follows the two tribes' uneasy path toward a new economy. They convert their lands from cattle forage to a rewilding ecosystem on behalf of giraffes, zebras, and rhinos—and tourists—while drought presses down on all.

I've asked ranchers whether they can see themselves in this role. They are rueful and disbelieving. I suggested to Gary Sprouse that research shows the land would be much healthier without cattle. How about a natural complement of wolves, tortoises, sage grouse, and other wild species instead? "That's the most stupid thing I ever heard of," he assured me. "So what good is the land, if it's not going to be productive? Cattle is what feeds people!"

So from my perch on their living room couch in a remote corner of the Arizona veldt, I held off asking these Mormon ranchers, the Bundys, to compare themselves to Ovahimba tribal folk. But I was fatuous enough, just the same, to ask whether they could envision a somewhat similar future. Instead of grazing cattle on public land, they might stay on here as stewards of a recovering landscape and its wildlife—jaguars and wolves and bighorn sheep, for example—and be paid to do it. I sold the dream pretty hard. They were gracious, but to no one's surprise the response was an unhesitant thumbs-down. There was a short laugh from Bill. "It wouldn't strike me very good," Orvel said. "No, I wouldn't want to see that. Money's not everything. . . . I'm not interested in taking care of somebody's wildlife."

Perhaps the long sale here would be religious, not ecological, and certainly not political. "We believe the land is here for our use," Bill Bundy said. "We are to better ourselves by using the land. The animals are here for our use. They were created, in our belief, by a loving Heavenly Father to better ourselves to prepare to return to live with Him. So I think if we're not taking care of the land, then we're not fulfilling our stewardship as children of God."

A religious person might speculate: what if the stewardship were redefined by some heavenly authority in an office in Salt Lake City, or the Vatican, to mean saying good-bye to the destructive cattle? I guess a revised catechism from a somewhat evolved deity might encourage disciples to ensure the survival, instead, of the earthly paradise of plant and animal species that once thrived here and are now all but obliterated.

Near the end of our visit the proprietor of a ranch not far from here, just over the state line near Bunkerville, Nevada, was mentioned. You may have heard of Cliven Bundy, a man of some notoriety. As it happens, he is Bill's cousin and Orvel's nephew. His herd illegally grazed a thousand square miles of federal public land on that very afternoon.

Cliven Bundy is in jail at this writing, along with two of his sons, Ryan and Ammon, awaiting trial. The two younger Bundys were principals in the armed occupation of the Malheur Wildlife Refuge in eastern Oregon, in early 2016. They had demanded that the federal refuge be relinquished to the local county government (which wasn't interested in becoming the new landlord, in any case). The occupation prominently featured lots of heavy weaponry and a cache of explosives and lasted for nearly six weeks. A sign posted by the occupiers read,

"BLM Another Intrusive Tyranical Government Entity Doing What They Do Best ABUSING POWER & Oppressing the Backbone of America."

The Bundy sons, with an entourage that included a rancher named Lavoy Finicum, left the refuge for a speaking engagement in another county. They were stopped on the road by federal officers, and Finicum was killed while reaching into his coat during a confrontation in which the others surrendered. Cliven Bundy, planning to join the Malheur group, was instead apprehended at the Portland airport. That effectively ended the occupation, but the Bundys have become iconic figures now to an unknown number of followers of various anti-federal-government groups. Finicum, who ran cattle on about twenty-five square miles of public land at the northern edge of the Parashant, is a martyr for them. Ryan and Ammon Bundy were later acquitted of charges in the Malheur occupation. The sole juror who spoke to the media—on the condition that his name not be revealed—said that in the Malheur trial, the not-guilty decision wasn't about the politics of the accused Bundys and their entourage. It was about the ineptitude of the federal prosecutors.

The fascinations of looming violence, angry speeches, big guns, and abundant cowboy regalia have consistently distracted the news media from detailed reporting about Cliven Bundy's tactics and their consequences for land management. They are worth recounting, as they are becoming more common.

For decades Bundy's herds wrecked watercourses, damaged archaeological sites, trampled critical desert tortoise habitat, and created highway hazards. They crossed over the state line onto the Parashant and wandered onto the Lake Mead National Recreation Area, far from the small federal allotments that he once grazed legally. At the time he lost his permit in the early 1990s, it allowed for only one or two dozen of Bundy's cattle to graze. By 2012 BLM surveyors found two thousand. For a quarter of a century Bundy did not pay grazing fees he owes to the U.S. government that total something over $1.1 million. "I don't recognize the United States government as even existing," he has often explained.

According to court records, the BLM sent Bundy a trespass notice and order to remove on July 13, 1993, and set a timetable for him to retrieve the cows. It expired. In late 1997 Bundy refused to meet with BLM officials, who sought to "resolve the dispute." In 1998 the matter wound up in federal court in Las Vegas, which affirmed that "the United States owns the Allotment where Bundy is grazing livestock without authority." It cited legal foundation for that ruling that goes back as far as the Treaty of Guadalupe Hidalgo between the United States and Mexico which, in 1848, ceded control of the land in question to the U.S. government.

So Bundy was ordered by the court once again to get his cattle off the allotment before the end of the year. The cattle stayed. A year further on, the court issued another order directing Bundy to comply. He didn't. Nervous federal authorities told the BLM to leave the cattle alone while they pondered other options.

"Notwithstanding the Court's Orders, Bundy continues to graze his cattle on the Allotment," the court observed fourteen years later, in July 2013. You wonder what the judge was really thinking, at that point, about the odd impotence of federal courts and federal law. Bundy was directed to remove his livestock again, and if they weren't the government was ordered to seize and impound the animals. Bundy was also instructed by the court not to physically interfere with the removal.

Nearly another year passed before federal officials, reportedly including everyone up to the secretary of the interior, decided to make their move, but in April 2014 the roundup began. "These people are thieves," Bundy told the Associated Press. "I'm going to fight for the Constitution and state sovereignty."

Fox News was an instant, overt supporter. "Big Government against the Rancher" was the tagline, as indignant Fox News-readers adrenalized the play-by-play on camera. Sean Hannity, who spotlighted Bundy continually, posed a rhetorical question during a live interview: "Why does the federal government need all that land?" Then Hannity figured things out for his audience: "You're using the public land.... I would think the government might be thankful, because you're cutting the lawn for free." He said the government was charging Bundy "huge amounts of money, right?"

Insurrectionists arrived at the ranch, in accordance with Bundy's plans. About four hundred self-styled patriots, dozens bearing sidearms and heavy weapons, came to the Bunkerville, Nevada, ranch from all over the United States. The media visuals were impressive: cowboys on horses with big U.S. flags, sonorous speeches by Bundy and his allies.

"Evidently, here in America, we don't actually own the property anymore, if we ever did," Scott Shaw, a member of the paramilitary Oklahoma Militia, explained on an Internet video that boasted of fifty thousand volunteers, at least some of them already on the road. "The Oklahoma Militia is ready to take up arms, if needed," the voice-over continued. Shaw warned the federal government, "You can do this legally, or if you want to do a land grab violently, you can do that. We're going to resist you."

Las Vegas police and Clark County deputies had been sent out as a "buffer" between the federals and the locals. "That was probably the scariest time in my life," police sergeant Tom Jenkins told an interviewer afterward. "When we first got out there ... that's all you saw. You saw the kids, and you saw the women, and you saw the horses in the background, and then you saw the men with guns, everywhere. They were laying on the ground, they were in the back of pickup trucks."

It would be wrong to conflate most public-lands ranchers with the unknown number of Cliven Bundys among them. A ranchers' group, while lashing out at environmentalists and their alleged federal co-conspirators, wanted nothing to do with Bundy's tactics: "Nevada Cattle men's Association does not condone actions that are outside the law in which citizens take the law into their own hands," the group announced.

But an array of politicians were more impressionable, courting sympathy for Bundy during the standoff. They included the governor of Nevada, a Nevada senator, many state legislators, and several political leaders who would later become Republican presidential candidates. They weighed in against the use of federal authority without mentioning any culpability on Bundy's part and without taking the federal court rulings, the laws, or the two decades of delay and defiance into account. They also downplayed or ignored the mortal threats posed to federal employees, who were pursuing their duty on behalf of all of us. At the ranch just then, snipers were training weapons on both local and federal personnel. Away from the cameras, law officers were taunted, threatened with rifles, and told to get ready to die.

Richard Mack, a former Arizona county sheriff and a leader of a group called the Oath Keepers, helped formulate a media-savvy plan. "We were actually strategizing to put all the women up at the front," Mack told Fox News. "If they're going to start shooting, it's going to be women that are televised all across the world, getting shot by these rogue federal officers." The tactic was reminiscent of terrorists who site mobile missile launchers in civilian neighborhoods. That way, any military response looks like an atrocity rather than a defensive measure. Meanwhile, the "rogue" federal officers were actually being guided from the top levels of the U.S. Department of the Interior. They were still trying to respond to a federal judge, a court order, and a growing, armed mob.

The federal government has not released a description or an after-action report on what went on that day and declined my requests for one. According to the account of a man named Ryan Payne (later arrested at Malheur), who stayed off-camera, he deployed his forces while Bundy presented televised demands to the BLM: open up all restricted public lands, remove BLM equipment, return the cows, and disarm federal agents. "We want those arms delivered right here under these flags in one hour," he demanded.

An hour or so passed. Then, suddenly, Bundy ordered his followers to block Interstate 15. After that he led the charge to release his penned cattle: "All we got to do is open those gates and let them back on the river!" he exhorted the crowd. As a final note he offered, "We're about to take this country back by force."

Law enforcement officials were outmaneuvered and probably out-gunned. Bundy's snipers, according to Payne, "had great lines of fire" with "overwhelming tactical superiority." The BLM's thinking at that point has not been clarified, but all armed federal officers suddenly withdrew, "because of our serious concern about the safety of employees and members of the public."

A few weeks later Bundy would confide to a political gathering in Saint George that God had inspired his battle. "The Lord told me ... if [the local sheriff doesn't] take away these arms from federal agents, we the people will have to face these arms in a civil war.... He said, 'This is your chance to straighten this thing up.'" So this Bundy, too, said his ranching was a divine mission, but in his case God told him, apparently, to perform his political beliefs by threatening violence. At the Malheur refuge, Ammon Bundy and several of his followers would also say they were divinely inspired.

Emboldened insurgents have worked hard to spread the muddle of Bunkerville doctrine—mostly in telegenic ways, so far, during the Malheur episode. Other post-Bunkerville events include these:

- Illegally grazing cattle on federal land is now called "going Bundy." Ranchers near Battle Mountain, Nevada, released hundreds of hungry cattle onto drought-stricken land that had been closed by the BLM. Its forage had already been devastated by grazing.

- Two Oregon men with plans to work a mine on public land near Grants Pass disputed their need to follow federal regulations. While waiting for their day in court, they called in the armed Oath Keepers militia to fend off any government intervention. Things quickly deteriorated. Miner Rick Barclay had to plead with his instant militia: "Please stop calling the BLM and threatening their personnel."

- The road into Recapture Canyon, Utah, had been closed to motor vehicles—not hikers or horseback riders—by the BLM after two men blazed an illegal seven-mile-long off-road-vehicle trail that vandalized Ancestral Puebloan archaeological sites. About 2,800 miles of other trails are open to those vehicles in southeast Utah, the agency pointed out. But dozens of all-terrain-vehicle riders, in a celebratory mood and led by county commissioner Phil Lyman, defied the closing and drove into the canyon anyway. Many of the riders were visibly armed. Ryan and Ammon Bundy and some of the Bunkerville militia showed up too and bulled their off-road vehicles into the archaeological areas.

"If things don't change, it's not long before shots will be fired," Lyman said later. "We can avoid it. But it's not going to be by the people changing their attitudes and accepting more intrusion into their lives. It's going to be by the federal government acknowledging people's freedom." Lyman and one other rider were later prosecuted, but the commissioner said he'd do it again, and Utah's state officials mulled over whether to cover his expenses for an appeal to a higher court.

After Bunkerville a survey of local law enforcement officials was undertaken for the Triangle Center on Terrorism and Homeland Security, by the sociologist Charles Kurzman of the University of North Carolina and terrorism expert David Schanzer of Duke University, funded by the U.S. Department of Justice. It found that law enforcement agencies think antigovernment extremists, not Muslim radicals, present the largest threat of political violence in the United States: 74 percent called antigovernment extremism one of the top three terrorist threats in their jurisdictions.

In the midst of all that potential mayhem, and despite the high stakes, we can celebrate that this is America after all, so even the pitched battle over public lands has its funny moments. When the Malheur occupiers sent out a call for sympathizers to mail in supplies, they received an abundance of nail polish, glitter, sex toys, and a bag of candy penises. An irate Jon Ritzheimer—who had organized anti-Muslim rallies in Phoenix and was seen near a mosque wearing a "Fuck Islam" T-shirt—complained that people were sending these boxes of "hate mail" to Malheur.

The cleverly produced country music video "Big Park," from an old Fox News program, has been a hit at "land rights" rallies. A truckload of strutting, singing park rangers invades the rural paradise of a rocking-chair grandma, with grandpa in overalls, and a mom and her angelic daughter. They're all evicted to make way for a national park "for the good of all," tied up, and hauled away in a livestock trailer. Tuneful: *We don't answer to the taxpayers / Or to your congressman / We don't take no from anyone / We just want to take your land.... Give the animals another place to play!... We'll say good-bye to humankind and build the Big Park of our dreams.*

Insurrection is not new to the United States. Daniel Shays led thousands of pitchfork-wielding rebels against the state of Massachusetts because it raised land taxes to pay off war debt, during the pre-Constitution era when colonists rid themselves of British rule and tried out a "states' rights" confederation. The rebels tried to overwhelm the Springfield Arsenal to seize arms and ammunition, but the state militia fired on them and killed several. Federal troops, under Gen. George Washington's direction, scattered the remaining insurrectionists the next day. The episode helped spur the movement toward the creation of a stronger federal government under a new Constitution.

Five years later an excise tax on distilled spirits was proposed. Small farmers in western Pennsylvania threatened tax collectors, burned the home of one of them, and threatened secession. Six thousand gathered outside Pittsburgh to protest and talked of laying siege to a federal garrison. Washington sent emissaries to parlay, but they were told that the tax would be resisted "at all hazards."

Reluctantly, Washington, now president, issued a final warning that called the tax evaders "treasonable." He pointed out that the tax had been levied by an elected, legitimate government. He arrived to lead the troops in person. The uprising ended with this show of federal force and an offer of clemency. Washington pardoned two men who had been sentenced to death for their roles in the rebellion—the first use of this presidential power, also granted by the Constitution, in the nation's history.

In hindsight, many have suggested that Cliven Bundy could have been dealt with by using smarter strategies than a clueless weaponized roundup—by prosecuting him for tax evasion, for example, or by impounding his cattle at their point of sale. At Malheur, however, some delay at least allowed the government to play its hand more deftly.

In fact, the federal government had considered moving on Bundy as long as twenty years before, but decided not to. George Washington wrote, "Commotions of this sort, like snowballs, gather strength as they roll, if there is no opposition in the way to divide and crumble them." Cliven, Ryan, and Ammon Bundy faced trial on charges related to the Bunkerville episode, but the Malheur acquittals and the antigovern-

ment and antijudiciary inclinations of the Trump administration made
any prosecution speculative.

Back out at the Parashant ranch house, Bill Bundy told me he went to
the Bunkerville standoff with three of his grown children, as an
unarmed supporter of his cousin Cliven. I asked whether Cliven looks
at his situation in a reasonable way. Bill and Orvel agreed that he does.
What about the federal judge's ruling? "Maybe the judge was misinformed," Bill said.

What is the right thing to do when you lose in court, though, misinformed judge or not? "Well, are you going to put your tail between your
legs and run and cry?" Bill asked. "We stand up for our rights, because
we were given those rights when we were made a country. And then if
they're taken away, are we going to just walk away or are we going to
fight for what we believe is ours?"

I suggested that if everyone who doesn't like a court ruling opposes
it with violence, things are going to get pretty crazy. Is that kind of
anarchy what we want to see? He replied with his own question: "Or do
we want to become a communist country because the government has
total control?"

A different answer for the anarchy question arose, though, as the
Malheur occupation sputtered to its conclusion. A Facebook link,
"Shootwelfareranchers," was posted for the "Public Lands Hunt Club
Sports Team." It invited readers to shoot cattle on public lands "owned
by ranchers who have publicly renounced their grazing contracts
designed to provide a minimum of environmental protections." That
might have been dismissed as an ugly hoax, but a Washington County
commissioner sent an "urgent alert" to the Utah State Cattlemen's
Association, warning of the threat to both cattle and ranchers.

"Many of these ranchers are publicly known due to their actions and
associations with the Bundy Family of Nevada," the "Hunt Club" page
advised. "Shoot straight. Be safe. Have fun." The page quickly drew
hundreds of "likes."

CHAPTER ELEVEN

Treasure Maps

The Endangered Species Act has given many animals a reprieve from the pathologies of state wildlife management. But removing them from that protection means they go back to the tender mercies of the states.

David Mattson, wildlife ecologist

On the Grand Canyon region's political landscape, and throughout western public lands, partisans of feral horses and burros compete for forage and water with cattle ranchers and deer hunters; recreationists contend against oil and gas drillers. Federal agencies exercise at least nominal control over this "multiple-use" empire and describe their role as an attempt to deal fairly with the needs of all interests at the same time. Politically connected suitors are shuffled higher in the deck, though.

For other players in and out of government, the game is instead to try to plan for climate change and the survival of natural systems and to make these the overriding priorities. Those plans come at the challenge in a variety of ways, at different scales. Globally, for example, as in the book *Half Earth*, by Harvard's ecopolymath E. O. Wilson, which prescribes just that much sequestered land for wild species; humans get the rest. At the continental scale a series of regional "rewilding" initiatives map out migratory corridors to connect national parks and other protected landscapes south to north, from the Appalachians to the

Pacific states. This is the map of hopes drawn by, among others, a group called the Wildlands Network (see Map 7).

National forests and Bureau of Land Management lands in the Grand Canyon region could be combined into a national monument extending north and south of the park. Partially protected habitat for migratory and broad-ranging animals already in residence, such as goshawks and bighorn sheep, would expand by several millions of acres. That proposed Grand Canyon Heritage National Monument is the product of years of effort by scientists and conservation groups such as the Grand Canyon Wildlands Council (see Map 8).

Near Malta, Montana, the American Prairie Reserve is bringing bison and other wild species back to 4,700 square miles of linked corridors and open grasslands, using its BLM grazing allotments along with purchased private land. There are campaigns to protect beleaguered species and particular habitats, by groups such as the Center for Biological Diversity, the Western Watersheds Project, the Grand Canyon Trust, and the Friends of Gold Butte, Nevada. The wolves, condors, sage grouse, jaguars and grizzlies, the tortoises, pollinators, and seeps and springs all have their partisans.

If public lands remain in federal hands with the survival of natural systems as their priority, the Grand Canyon region could link south and east through the Mogollon Rim lands and into New Mexico, as shown on the map. To the north and east there are still expanses of public lands from the North Rim's Kaibab National Forest through a proposed Theodore Roosevelt Wildlife Conservation Area, to the Vermilion Cliffs National Monument, which is like an eastern bookend for the rectangular area called the Arizona Strip along the Utah border. Its western bookend is the Parashant. Continuing north are Grand Staircase-Escalante National Monument and Zion National Park. They connect, potentially, to the "spine of the continent"—the Rockies—and then farther north, all the way through the Yukon to northern Alaska.

But we are moving quickly, instead, along another route into the future of public lands, with an alternative map as its guiding vision. Its authors are high-decibel, well organized, and politically potent. Instead of combining public lands, it would atomize them. It marks out a different set of payoffs, too. This map leads particular private economic interests—gas and oil drilling, logging, grazing, mining, real estate development, and mechanized recreation   toward their goals. Healthier air, soil, and water, the survival of wild species, and protection for living habitat and ancient archaeological sites are no longer emphasized.

Thus President Trump announced, early on, a "review" of the protections, and the boundaries, afforded by twenty thousand square miles of national monuments, including the Parashant, Vermilion Cliffs, Grand Staircase-Escalante, Bears Ears, and more than a dozen others. The Republican Party has gone even further, however. It advocates dismembering the federal land heritage and transferring an unspecified portion of the fragments to the states. This antifederal agenda duplicates what the motley militias of the Bundy fringe offer. But it connects, instead, billionaire funders and their political interests with state and federal legislators. They regard the Bundy movement as a sideshow, sometimes applauded and at other times shunned.

Unlike the Bundy-style "sagebrush rebels," this campaign operates within the law. It has gained powerful momentum, though polls in the affected states show that it is still a minority view. It has had far better success in both state and national legislatures, where you hear this now as dirt-level basic doctrine: control over federal public lands rightfully belongs to the states. The Arizona legislature, for example, approved this bill in 2012: "On or before December 31, 2014, the United States shall extinguish title to all public lands in and transfer title to this State."

That would have included national forests, monuments, historic sites, archaeological sites, wilderness areas, and of course the national parks: Canyon De Chelly, Casa Grande Ruins, Chiricahua, Coronado, Fort Bowie, Glen Canyon, Lake Mead, Montezuma Castle, Organ Pipe, the Parashant, Petrified Forest, Saguaro, Sunset Crater, Walnut Canyon, Wupatki—and not least, Grand Canyon. The bill included a provision that the state had the right to sell off these new acquisitions as it chose. It was vetoed by the governor, a Republican.

A more common formula would exclude national parks, military bases, Native American reservations, and congressionally designated wilderness areas but hand everything else over to the states. Legislatures in Utah, Wyoming, New Mexico, Colorado, Nevada, and Idaho have explored and sometimes passed legislation demanding that the national government relinquish federal land. The Utah legislature authorized $3 million to be ready to sue the federal government over federal land management and $500,000 for a study of possible strategies to take control of forty-seven thousand square miles of federal public land. The Republican Party platform for the 2016 national election stated, "It is absurd to think that all that acreage must remain under the absentee ownership or management of official Washington. Congress shall immediately pass universal legislation providing for a timely and orderly mechanism requiring the federal government to convey certain federally controlled public lands to states."

This controversy has a long history. But what really drives the surge of interest in just the past few years? No other factors look as powerful as the alignment of stratospherically wealthy private interests with antigovernment ideology. Their campaigns are bankrolled by corporations and individuals that spend hundreds of millions of dollars on lobbying and political candidates—although the sources and amounts are now often hidden.

For example, those in the fossil fuel industry—including the Koch brothers, whom we've met earlier—are a potent source of antipathy to government regulations for health, safety, and the environment, including attempts to slow global warming. For them, a great deal is at stake in how—and whether—the natural systems on those public lands are managed or protected.

The BLM administers more than 23,000 active oil, gas, and coal leases on four hundred thousand square miles of surface land and controls about a million square miles of subsurface rights. The Bureau of

Ocean Energy Management supervises about 8,300 active oil and gas leases on the seabed of the Outer Continental Shelf. Together coal, oil, and natural gas produced on federal lands supply about 25 percent of the total fossil fuels the United States produces.

Federal rates for these leases are deeply gratifying, unless you're concerned about, say, the state of the U.S. Treasury. If leases were priced competitively, the government would earn hundreds of millions of dollars more each year, according to Jayni Hein of the Institute for Policy Integrity at the New York University School of Law. Energy companies enjoy rates that haven't kept up with inflation, or with fossil fuel prices, for decades. Oil may be cheaper or dearer as you read this, but prices tripled from 2003 to 2013, while the minimum the government charges for oil leases was unchanged.

In fact, for decades the minimum bid to lease an acre of public land for one year for potential fossil fuel production has been about what you pay for a cup of coffee: $1.50. Adjusting only for inflation since that price was set, long ago, would raise it to $24.00. Annual rental fees, which companies pay to hold and explore federal lands before production, are just as low. The royalty rate for oil and gas produced onshore has not changed, either, since 1920. It is far less than what individual states, and other countries, charge for the same uses of their public lands.

Calculating the real costs of these subsidized dream deals for industry would help. "Those bargain prices give private companies a windfall while depriving American taxpayers of a fair return from energy production. Instead, the public has been left to pay for many of the social and environmental costs of fossil fuel operations," Hein observes, including wildlife disruption, health impacts, and the ruin of former hiking and scenic areas—not to mention the carbon pollution that is accelerating climate change.

The secretary of the interior has a great deal to say about such matters, as the federal officer atop the chain of command of the National Park Service and the Bureau of Land Management. Congressman Ryan Zinke of Montana was appointed to that cabinet post by the Trump administration in 2017. Oil and gas interests contributed more than $300,000 to Zinke's one congressional campaign.

One of the wellsprings of agitation to let the states take over all this publicly owned fossil fuel wealth is Koch Industries, the second-largest private company in the United States, according to *Forbes*. It is owned by our ranching friends David and Charles Koch, both on the list of that magazine's ten wealthiest individuals on Earth. During the Obama years the brothers' personal fortunes nearly tripled, from $14 billion each to $41.6 billion each—though they were a bit off that peak more recently because of declining oil prices.

Koch Industries employs more than a hundred thousand people in sixty countries. Some of its primary holdings are in the oil and gas industries as well as timber, pulp, and paper. BLM records show that a Koch Industries subsidiary controls 221 oil and gas leases in six states.

Koch Industries' U.S. lobbying expenditures, second only to Exxon-Mobil among oil and gas interests, totaled $103 million during the most recently recorded decade (the figure for the whole industry was well over a billion dollars). The brothers and their company also contribute millions to support both state and national political candidates who favor their agenda. For the 2016 elections alone, each brother pledged $75 million to help reach a goal among their network of funders of $889 million, rivaling the budget of the Republican Party itself.

The Kochs have organized and helped fund dozens of groups—which detractors and supporters alike sometimes refer to as "The Kochtopus"—to campaign for a long list of political causes, including millions to undermine public confidence in the science of climate physics and its projections about global warming. It is fair to loosely characterize some of these initiatives as "libertarian" or perhaps just "antigovernment," and it is also true that many of them coincide well with Koch business interests.

The to-do list prominently includes fighting environmental protections on federal public lands and promoting their takeover by the states. That agenda is pushed by Koch-supported groups such as Americans for Prosperity, the American Lands Council, and the Property and

Environment Research Center. The Kochs are also funders of the American Legislative Exchange Council, which circulates "model" laws to friendly state legislatures and then pushes for their passage. One model bill, for example, states that global warming may lead to "possibly beneficial climatic changes. Further, a great deal of scientific uncertainty surrounds the nature of these prospective changes." These are the sly clichés of the long-standing fossil fuel industry disinformation campaign about climate disruption.

Yes, there will be benefits from warming, and their scale will be minuscule compared to the disruption it is already causing. Yes, there is divergence among scientists about the details of what will happen. That honest degree of uncertainty is part of the credibility of their reckoning. But the scientific consensus about projections that global warming is indeed underway, and that its consequences will be severe beyond calculation, is nearly universal.

Another model bill, the Disposal and Taxation of Public Lands Act runs to 3,700 words of lawyered radical change, but the gist is plain enough: "NOW, THEREFORE, BE IT RESOLVED that the Legislature of the state of [insert state] demands that the Federal Government immediately dispose of the public lands within [state]'s borders directly to the state of [state]."

The two trains of thought—consolidation versus fragmentation of our public estate—collided in 2010, when the Bureau of Land Management embarked on an internal discussion of a new management outlook and higher levels of protection for some areas. Robert Abbey was director of the bureau at the time and his boss, the secretary of the interior, had issued a call for new initiatives. A career-long agency manager in several western states including Arizona, Abbey felt some pride in the fact that the BLM is the single federal agency that returns more money to the government—mostly through oil and gas leases on its lands—than any other. He does not oppose drilling, mining, or grazing on public lands but, he told me, "conservation is part of the BLM's multiple-use mission. It isn't all about leasing areas for oil and gas. It isn't about just grazing cattle on public lands or mining. Conservation sometimes gets lost in that discussion."

So his office drafted a paper titled "Treasured Landscapes," which highlighted fourteen broad patches of public land. These represented special opportunities, and for urgent reasons: global warming, the paper argues, makes it imperative to see past the jurisdiction lines of the Forest Service, the BLM, or other agencies. These landscapes must be consolidated, protected, and connected with wildlife corridors. "The BLM hopes to participate fully in the effort, and help lead the charge," the document stated.

The plan also recommended a strong departure from routine, with major implications for federal land policy. It proposed that a new system be used to consider the value of ecosystem services on public lands, just as the value of mining, timber harvests, or cattle grazing are already quantified. The approach would "allow land-use decision makers to act with a fuller knowledge of the trade-offs involved in the choice to conserve an existing landscape, or permit new development."

The paper highlighted 219,000 square miles of BLM lands—about the size of Colorado and Wyoming combined—that have "unspoiled beauty" and a critical role in habitat conservation, as well as historical and scientific significance: "We now know that these large-scale ecosystems, watersheds, air sheds, and migratory pathways exist and function only at their natural scales, regardless of jurisdictional boundaries."

The plan withered under fierce political opposition, just as most of Interior Secretary Bruce Babbitt's agenda for a new BLM had faded as the Clinton era closed. Abbey left the BLM a couple of years later. The new national monuments created during the Obama administration included only three out of the fourteen that were on the "Treasured Landscapes" list. And many of those new Obama monuments also incorporated hundreds of grazing permits as well as oil, gas, and mining operations, which will all continue. The proposal to value ecosystem services on public lands receded from view too, though it still appears as a buzz phrase in some agency materials. "Treasured Land-scapes didn't materialize," a BLM spokesperson told me. "It's ancient history around here."

If federal public lands fell under the control of state authorities, con-servation efforts would vary, just as state wildlife agencies do. It depends to some extent on which political interests have won the most recent elections. This whipsaw makes long-range planning for natural lands and resources difficult, perhaps nearly impossible.

In many places the state agencies' interest in the environment as anything other than a conveyor belt to deliver fish and game to anglers and hunters seems meager. Nonetheless, federal land managers collab-orate with state wildlife agencies, because they think it a sensible divi-sion of labor, or politically expedient, or because their own funds and personnel are overstretched.

The Arizona Strip BLM district, to the north of Grand Canyon, includes the Vermilion Cliffs National Monument on the east and part of the Parashant on the west. One benchmark for the health of the plant cover on this huge federal area is "potential natural community"—the vegetation that would occupy a site over time if it were not influenced by people. But hunting mule deer is "the big ticket item here on the Strip," the BLM's lead rangeland-management specialist, Whit Bunting, told me. "That's what we are really famous for.... Really, to promote mule deer, we don't necessarily want to manage for that potential natural community." Who decides how many deer are too many? "That would be Arizona Game and Fish," he said. "That is really the call of the state agency."

The future for federal land managed by a state like Arizona is easy to discern in the public record. As noted earlier, there are no biologists, ecologists, or endangered species specialists on the Arizona Game and Fish Commission. Four out of five commissioners are sport hunters; three are NRA members. The commissioners were appointed by a governor who does not consider it the case that humans have a role in causing cli-mate change. Under that form of stewardship it is unlikely that natural habitat would be managed by the state in anticipation of global warming.

David Mattson, a former U.S. Geological Survey wildlife ecologist now at Yale University echoes other scientists I have spoken with. He says many state game and fish agencies are intransigent about the role of wildlife, especially predators. Website rhetoric to the contrary, in the eyes of those agencies, predators are often just blood sport trophies, competitors, or dispensable nuisances.

Climate change and invasive species are diminishing carrying capacity for grizzly bears and mountain lions, two species he has focused on, Mattson explains. "One of the key factors, though," he adds, "has to do with the nature of state wildlife management. The Endangered Species Act has given many animals a reprieve from the pathologies of state wildlife agencies. But removing them from that protection means they go back to the tender mercies of the states."

One of the BLM areas briefly enshrined on the now-obsolete Treasured Landscapes list was Gold Butte—560 square miles of baking red sandstone and gravel washes, a habitat for bighorn sheep, banded Gila monsters, great horned owls, Joshua trees, yucca forests, and desert tortoises.

Former Secretary of the Interior Bruce Babbitt, whose forebears were Arizona cattle ranchers, once summed up the research literature as it applies to places like this: "I am now convinced that livestock do not belong in arid deserts.... I am here to say the presumption that grazing is the dominant use of our public lands is the artifact of a distant past and must be replaced."

A boom camp materialized here briefly in the early 1900s for about a thousand gold miners. It was one of a series of flickering hopes for mica, magnesite, vermiculite, or copper miners, dating back to the Spanish reconnaissance of the 1700s. Three thousand years of human history precede that, still to be seen in rock art, caves, agave roasting pits, and ancient campsites.

The Gold Butte area lies along Interstate 15 about eighty miles northeast of Las Vegas, on the western edge of the Parashant. It is next to the route through southern Utah and Paiute country that heads up to the North Rim of Grand Canyon. One of the rare highway off-ramps in this remote area is for Bunkerville, Nevada. The Cliven Bundy ranch occupies part of Gold Butte.

The morose condition of much of our national land heritage is not widely known. It's an obscure outback, except to industrial interests, hikers and environmental groups. There's little in the way of public narrative yet about why these lands matter—except to oil, gas, and grazing interests. President Barack Obama pursued what he called an "all of the above" strategy during his two terms, fighting climate change with some policies while, paradoxically, promoting more gas, coal, and oil development on public lands with bargain-basement prices and lax regulation of pollution standards.

But in his final State of the Union message, he signaled a shift: "Now we've got to accelerate the transition away from old, dirtier energy sources.... That's why I'm going to push to change the way we manage our oil and coal resources, so that they better reflect the costs they impose on taxpayers and our planet," he said.

Assessing true costs is a start. That would make it easier for us to sense the magnitude of what is at risk. We could see the designs of the Clivens and the Kochs more clearly, in terms of their consequences. Thanks to the long, illegal residence of Bundy's two-thousand-head cattle herd, some of Gold Butte is, as one biologist put it, "beaten to shit."

As he left office, Obama designated Gold Butte and the exquisite Bears Ears region in Utah—their wildlife, history, archaeological sites, and scenery—as new national monuments. Trump-era successors arose with new confidence to seek repudiation of public lands protections.

Like the rest of the United States' public lands, this is contested territory. Whose treasure map will it be part of?

CHAPTER TWELVE

Thrill Rides

*My hike was virtually never free of the incessant drone
of aircraft.*

    John McCain, Arizona senator

Less than three miles outside the South Rim entrance station, near
Tusayan, you can watch an enduring spectacle at Grand Canyon Airport. Amid deafening thap-thap-thaps and gale-force downdrafts, helicopters in festive livery lift and land from a line of seven pads—one
every few minutes, when I visited. A river of aircraft flows out over the
Canyon: some fifty thousand flights a year just from this location and
double that number of low-altitude aircraft from Las Vegas and other
points of origin, ferrying more than 423,000 tourists and an unknown
number of others.

    The biggest operator here, Papillon, bills itself as "The World's Larg-
est Aerial Sightseeing Company" and "The Only Way to Tour the Grand
Canyon." The company recently fanfared a new $9 million headquarters
near Las Vegas, where many Canyon air-tour flights originate. This for-
midably busy scene has at least a couple of backstories: how helicopters,
airplanes, snowmobiles, and all-terrain vehicles are faring at national
parks—the United States' refuges for "nature, unimpaired"—and on wilderness areas, national forests, and other public lands. Behind that tale is
another, it turns out, about our national system of politicking.

    After listening to the ponderous 'copters for a while, I headed back
out to the Canyon rim. The Pima Point overlook on one of the main

park roads has been closed to auto traffic except for shuttles, so quiet reigns there at times. It's one of the viewpoints from which you can spy, far below between the towering mesas, a slice of the sinuous green Colorado.

And I could hear that flow as well. The delicate susurrus of Granite Rapids rises a couple of miles from the hot Canyon bottom all the way up to this breezy plateau, to me and a couple of pinyon jays trading calls in the trees. It was a beckoning, ciphered signal I thought I might decode, if I could just listen long enough. Then a helicopter motored past, a mile or two out over the Canyon—radiating about thirty-five decibels of noise at this distance.

River, jays, and breezes were smothered within the insistent mechanical pulse. Natural sound is fugitive. Just three decibels of noise—less than 10 percent of the 'copter's—overrides birdsong so strongly that if you were listening a hundred feet away, you'd have to move fifty feet closer to hear it again. Of course, my distracted gaze followed the big wasp too and so, I suspect, would yours.

Republican John McCain, senator of Arizona, was once subjected to that same racket when he hiked down the South Kaibab Trail to embark on a raft trip, and "the privilege to experience, first hand, the beauty and grandeur that is the Canyon." But as the freshman senator told a committee hearing thirty years ago, "My hike was virtually never free of the incessant drone of aircraft. Following that experience, I resolved to do something about this problem."

His frustration then is widely shared now. Ten national parks are trying to manage high-intensity overflight noise problems. A hundred or so other parks have air-tour applications under consideration. But the problem is broader still. Aircraft are only one form of disruptive technology, lucrative for promoters but inimical to both wildlife and the visitor experience.

Several dozen national park units now allow snowmobiles, for example, though in 2000 the Park Service had planned to phase them out nearly everywhere. Instead, a new national administration took office, and that policy was countermanded. At Yellowstone more than a thousand snowmobiles a day were okayed (the number has been reduced since then).

Nearly all national forests and BLM lands and a dozen national park units allow recreational all-terrain or four-wheel-drive vehicles on at least some of their landscapes. A long series of studies have documented their damage where the use is not carefully monitored and regulated. Erosion, soil compaction, and destroyed vegetation are "obvious to even casual observers," as one overview of the research notes, summarizing the phenomenon simply as "environmental degradation."

The vehicles create mud holes and gullies that intensify erosion, spread invasive species, and load mud and sediment into streams, which threatens fish and insect populations. Wildlife are negatively impacted by the presence and noise of ATVs and snowmobiles, and snow compaction affects the survival of small mammals in winter. The machines are sometimes used to pursue wildlife to exhaustion, and their use enables vandalism in areas formerly protected by their remoteness.

But more than 80 percent of BLM and Forest Service field personnel surveyed by the Government Accountability Office report that they don't have enough resources to enforce regulation of off-road machines, and they called the problem a "great challenge" in their work. Half the national park officials surveyed said the same thing. They were also asked if they could "maintain existing off-highway vehicle areas in a sustainable manner," which includes compliance with regulations. Half the Forest Service officials replied, "Definitely not," as did 79 percent of the BLM and 38 percent of the Park Service respondents.

Despite its "Green Parks" commitments to promote sustainability, the Park Service itself is a car junkie at times, remarkably unready to consider moving them out of the parks when opportunities arise. At Glacier National Park, whose namesake glaciers have all but disappeared due to global warming, a staggering $165 million is being spent to patch up the fifty-mile-long Going-to-the-Sun Road, a multiyear project due to wrap up in 2017. The road is a national treasure of some sort, if you subscribe to Park Service rhetoric: an "engineering marvel," and so on. Dyspeptic view: in the United States roads are commonplace, but protected alpine ecosystems are an extreme rarity. This road allows a half-million cars to motor on through each year, emitting ten thousand tons of carbon dioxide into our climate greenhouse.

Carbon dioxide or not, the view from inside your car is skimpy, and it's not quite safe to look out there anyway. The driving's all stop-go, with tight curves, tight clearances, and traffic congestion that all demand full concentration. The main halt, the big Logan Pass parking lot, is choked full nearly all summer long, according to park management, despite a recent expansion—so are the turnouts that are supposed to enable you to emerge for a rushed glance at panoramas that call for long contemplation. I know sanctimony's the worst cologne, but consider: as a passenger on a safe, quiet, comparatively cheap, nonpolluting electric rail shuttle over the pass, you could really enjoy the view. With the $165 million road repair money in your jeans, you could get that done.

Rick Ernenwein worked for just about half of his thirty-five-year career on noise-and-overflight issues at several national parks, but mostly at Grand Canyon. He was a seasonal ranger there in the 1970s, when air tours were a high-hazard carnival, occasionally joined by rogue military pilots screaming down the inner gorge. One flew a jet under a low suspension bridge on the Canyon bottom. Commercial air-tour pilots, often flying Korean War-era single-engine planes, earned tips from passengers with daredevil rim dives, by shooting the notches in Canyon formations or with close buzzes of river rafters. Twice, out on the trail, Ernenwein felt he had to duck under the swoops of very low-flying planes.

"Probably the thing that I remember most from my first years at the Grand Canyon was how blue the sky was," he told me. "I mean, it was just a lot bluer than I had experienced. It was pristine air. That gradually eroded over the years, with air pollution creeping in. It was kind of the same with the noise."

Noise was already on the Park Service radar, you might say, long before that. And in a five-year period in the early 1980s there were ten air crashes and fifty-three deaths involving tour aircraft in and around Grand Canyon. That, and his noisy raft trip, got Senator McCain's attention. Urging passage of a bill intended to restore "natural quiet" to most of the park, he made a few points with ringing clarity.

"I would like to make it clear that, in my view, the Grand Canyon does not exist for anyone's financial benefit," he said, adding later that "when it comes to a choice between the interests of our park system and those who profit from it, without a doubt, the interests of the land must come first." Not only this park, but the entire national park system would be jeopardized, "if we fail to take the steps necessary to retain the fundamental essence of the Grand Canyon, its beauty, its splendor, the timelessness and the solitude."

The Park Service, McCain concluded, has the expertise and the authority to regulate aircraft noise levels at Grand Canyon, and both Congress and the Federal Aviation Administration should butt out. The FAA had shown little interest in noise impacts on either visitors or wildlife. "Then John McCain stepped in and really kind of led the charge," Ernenwein told me.

Though he was not opposed to air tours, McCain's priorities did not please the industry, which complained that the bill would cripple them financially. It became law in 1987. McCain won awards from environmental groups, some of whom saw a fine opportunity to rid the Canyon of the hated overflights. They pushed what they called the "Quiet Canyon" proposal: no aircraft there at all.

Given park budgets and the Gordian-knot intricacy of the research, we don't know much about the impacts of loud, frequent, unnatural noise on the true inhabitants of the Canyon: the condors, falcons, and owls; the ringtails and badgers and dozens of other species. We do know that endangered peregrine falcons suffer "major adverse impacts in areas beneath air-tour routes."

One comprehensive study examined how 138 different species of birds in North America, Europe, and the Caribbean change their behavior in response to nonnatural noise. They avoid it. Their natural "acoustical environment" often cues their feeding, breeding, and survival. Overall, the research found that human-made noise is "a powerful sensory pollutant" that interferes with birds' abilities to send, receive, and respond to sound signals in their habitats.

Another study found that overflights disturb the grazing patterns of Grand Canyon's desert bighorn sheep—enough to make them work harder for less food. They are an extremely rare, genetically pure population unmixed with other sheep species. There is no money to closely study noise impacts on bighorn sheep or on dozens of other kinds of animals.

Among us humans, natural quiet is reported as "extremely important" by huge percentages of both day-trippers and back-country trekkers at the park. Echoing many other such surveys, one found that 72 percent of visitors named "Providing opportunities to experience natural peace and the sounds of nature" as one of the most important reasons to visit a national park.

A third of visitors on short hikes at Grand Canyon report "moderate to extreme annoyance" when exposed to aircraft noise. Among those who visit only louder places—developed areas and overlooks—four out of ten nonetheless say air tours are "inappropriate." Hikers and river runners agree, and far more strongly, surveys have found. Except for its small but heavily trafficked rim enclaves, Grand Canyon is supposed to be managed, after all, as a wilderness area—with "outstanding opportunities for solitude."

Research has long since confirmed that noise can induce stress, annoyance, distraction, and declines in mental performance. Three cognitive psychologists and a neuroscientist recently posed this question: can human-caused noise in national parks impair memory? Their research finding: yes. Subjects were shown scenic natural images while listening to interpretive talks about them, for example, and noise significantly affected their ability to remember the information.

That foundation of plain logic was the basis of the McCain legislation. "Noise associated with aircraft overflights at the Grand Canyon National Park is causing a significant adverse effect on the natural quiet and experience of the park," the law stated, and it gave the NPS and the FAA just three months to resolve safety issues and "provide for substantial restoration of the natural quiet." The safety issues were addressed in short order.

To address the noise problem, however, the government embarked on what would become a very long journey indeed. It included litigation from both environmental groups and the air-tour industry. More important was what looked to some Park Service personnel like steadfast opposition from the FAA.

The FAA, critics said, used alleged safety issues to trump Park Service attempts to diminish noise. At one public meeting, Ernenwein recalled, an FAA regional administrator stood up to say that there was a constitutional right to fly anywhere, unless there were safety or national security concerns. "The FAA couldn't believe that any rational person could have any problem with the aircraft noise levels at Grand Canyon. They were used to dealing with noise levels around airports that could almost make a person's ears bleed," Ernenwein said. An FAA spokesperson told me that the agency cooperated consistently with the Park Service to achieve the goals of the act, however.

McCain's legislation required the Park Service to survey aircraft noise impacts at all national parks. When that survey appeared, its preamble was well outside the norm for government reports. The essayist Pico Iyer was cited: "Silence is something more than just a pause; it is that enchanted place where space is cleared and time is stayed and the horizon itself expands. In silence, we often say, we can hear ourselves think; but what is truer is that in silence we can hear ourselves not think, and so sink below ourselves into a place far deeper than mere thought allows. In silence, we might better say, we can hear someone else think." The judgment of Herman Melville was included, too: "All profound things and emotions of things are preceded and attended by Silence.... Silence is the general consecration of the universe.... Silence is the only Voice of our God."

There were other gods astride Grand Canyon though, as a decade passed and McCain's noise legislation was mostly unimplemented. And though their revenues had quadrupled in the meantime, the air-tour industry had not been appeased. Nationally, the U.S. Air Tour Association chafed at the prospect of federal regulation to deal with what seemed to them a nonissue at national parks.

The association's president and its chief lobbyist met with leaders from the American Recreation Coalition, which represents the motorcycle, snowmobile, recreational vehicle, and ATV manufacturers, among others. They discussed the threat of the Park Service's interest in natural quiet as a resource to be protected. "The issue of noise at national parks is not limited just to the air tour industry," USATA lobbyist James Santini warned. "The evidence is clear that the NPS grand plan is to eliminate anything from parks which produces any level of noise above a whisper." He called McCain a "nemesis," but promised that USATA was working on changing his mind.

By 2000, thirteen years after the act's passage, the noise issue was still not resolved. More interagency meetings were held. But the zeitgeist had changed at the White House, and elsewhere. One political appointee of the new administration was Paul Hoffman, a former director of the chamber of commerce in Cody, Wyoming, who had battled for more snowmobiles at Yellowstone. He had also opposed the reintroduction of wolves there, which he characterized as "the equivalent of detonating a nuclear bomb in the West."

Hoffman was named the Department of the Interior's deputy assistant secretary for Fish, Wildlife and Parks, which ranked him above the director of the National Park Service. He concerned himself to overrule the decision of a Grand Canyon superintendent to remove religious

plaques on view near the South Rim and ordered park officials to continue sales of a book that espouses the geological theory that Noah's
flood created the Canyon six thousand years ago.

In 2005 Hoffman sent down a proposed revision of the Park Service's
book of basic management policies. That may sound like a ho-hum
event, but this is the document that translates existing law into day-to-
day action for park superintendents and program managers, according
to Bill Wade, the chair of the Coalition of National Park Service Retirees and former superintendent of Shenandoah National Park.

A sentence from the existing policy that was deleted by Hoffman:
"the Service will strive to preserve or restore the natural quiet and natural sounds associated with the physical and biological resources of the
parks." Also cut was a reference to natural sounds like "waves breaking
on the shore, the roar of a river or the call of a loon." Hoffman's draft
stated, "the variety of motorized equipment and mechanized modes of
travel are diverse, and improved technology … makes their use more
feasible in larger areas of the parks." It noted that some uses of these
vehicles "may cause impairment of resources or values," but it proposed
changes that invited more off-road use of snowmobiles and all-terrain
vehicles in all the parks.

The Hoffman alterations would also have promoted more cell phone
towers and more low-flying tour planes and would have taken away
administrators' use of the Clean Water Act and the Clean Air Act to
limit new development within the parks. Snowmobiles would be
allowed to travel over any paved road in any national park. Grazing and
mining were redefined as "park purposes," and air-quality standards
were lowered. The proposals would have made it more difficult to create new national parks and would have eliminated nearly all references
to evolution in park resource and education programs.

"Impairment" of the parks' natural systems was redefined so that
only actions that could be proven to cause "permanent and irreversible"
damage were prohibited. Under that logic, Wade said, "I suppose you

could do anything to a wildlife population as long as you preserved the last breeding pair."

It would be a mistake, he told me, to attribute all this just to Hoffman. His work had powerful political support, all the way up to the secretary of the interior. "We had no doubt they were serious about making these changes," Wade said. A Park Service work group discussed the draft for three days, anxious about how much of its damage they could mitigate without falling afoul of their bosses at the Interior Department. "I was profoundly shocked at how far it went," a participant in the workshop told a reporter.

Media coverage and public arousal, opposition within the Park Service and from groups like Wade's, forced the administration to walk back the attempted coup. Fortunately, that's all ancient history and no longer of concern? "I would challenge that," Wade told me, awhile before the 2016 elections. "I think it could happen again, at any time."

The Park Service's willingness to compromise on the noise issue was on display well before Hoffman's arrival. The option to ban all tourist overflights was repeatedly rejected. Despite their mission to protect nature and the visitor's experience of nature, park administrators reached the official conclusion that aircraft are an "aerial viewing platform" with a legitimate role in conveying tourists. (Strangely, other "aerial viewing platforms," far less intrusive, are appropriately disallowed at Grand Canyon. Hang gliders, for example. Drones with video cameras, too.)

Next, the Park Service yielded to the FAA on addressing the noise from high-altitude overflights by commercial jets and other aircraft. Research shows that these high flyers make much, at times most, of the noise in terms of the length of time they are audible. Monitoring even at remote backcountry sites found that high-altitude jets were audible for 30 to 40 percent of the day. The FAA had argued unpersuasively in two courtroom battles with environmentalists that commercial jet noise in Zion National Park and at Grand Canyon should not be included in noise calculations because it was trivial. The FAA lost. Jet noise is loud and pervasive.

Nonetheless, those airliners were now off the table, as the Park Service dropped the argument. In exchange, the FAA "agreed to consider reducing aircraft noise over the park in the future." (A decade later, no progress had been announced by the FAA on that project.)

The Park Service's interpretation of the act's language—"substantial restoration of natural quiet"—was always a moving target in the negotiations, and it moved again. Now that high-altitude planes were untouchable, by 2008 "quiet" had become this formulation: noise generated by air-tour planes and helicopters should not be audible in at least half of the park, for at least three-fourths of each twelve-hour daytime day.

So as much as half of Grand Canyon had been ceded to air-tour noise, and even the "quiet" half could also be noisy as much as a quarter of the time. Once the "no tourist overflights" proposal disappeared, nearly any definition of "substantial restoration of natural quiet" could be put forward and rationalized. It was far too generous a concession, if you listened to environmental groups. As for the industry, it still felt burdened.

Meanwhile, a series of legislative maneuvers eliminated more and more classes of noise-generating overflights from the negotiations altogether. Sen. Harry Reid, Democrat, added a rider to a bill that defined air-tour flights from Las Vegas over the Canyon as exempted "transportation." The Hualapai tribe was granted a temporary exemption for its own helicopter fleet, still in place twenty years later (the Navajo tribe is trying to get in on the overflight action now too). Refueling flights, training flights, flights into the park from the Lake Mead corridor—altogether, these exemptions more than double the actual number of low-altitude overflights to well over a hundred thousand a year, only half of which were even under consideration for noise regulation.

Research continued at the park. Along with planner Ernenwein, specialists Sarah Falzarano and Laura Levy gathered field data, refining what the Park Service knew about noise in the Canyon. They analyzed sounds gathered by equipment posted in remote areas, designed to try to capture and measure "natural quiet." Often the monitors were accessible only via river rafts or rugged hikes. "I was hearing things a lot of people don't get to experience," Levy told me. "Birds in the morning, owls at night, elk bugling—eerie sounds." The microphones picked up distant thunder and nearby breezes and, on one occasion, the crushing, grinding roar of a flash flood, within a ravine that Levy had hiked out of only a couple of hours before.

Falzarano still remembers the recorded sounds of coyotes, gathered somewhere in the rocks: "It transports you to the night, and the desert," she told me. There were also recurrent gurgles and the smack of elk slobber, because the huge animals became fond of licking and chewing the foam-rubber covers on the microphones. And there was aircraft noise, even in backcountry areas expected to be entirely free of it.

Their supervisor, Ken McMullen, knew that during all this meticulous labor the air-tour industry was also at work. "The question was, how hard can we push, before the politics intervene?" McMullen told me. "We wanted to get as much as we could for the benefit of the people and the resources in the park, but we all knew we had constraints. Sometimes it kind of made me think, well, what am I doing? I still wanted to do a good job," he said, and base decisions on well-planned and well-executed soundscape studies, given the limits on research time and money. "But I realized it was likely that things were not going to change a whole lot, and that the Park Service would be pressured into compromising some of their own values even further."

By 2012 a seven-hundred-page Environmental Impact Statement (EIS)—the noise plan meant to finally resolve the overflight issue—was circulating for public comment. It represented millions of tax dollars, decades of Park Service research, and an elaborate public-involvement process—in a way, the culmination of McCain's first noise initiative, at long last. But contradicting his dictum that national parks don't exist to make money for anybody, the Park Service's own EIS now stated that an "economically viable" air-tour industry was one goal for its Grand Canyon noise plan. We were a long way from the early days, when private interests who wanted to own trails and levy tolls on hikers or to homestead, mine, and farm within the park were handed their walking papers after protracted legal battles. Then, the park was considered nearly sacrosanct.

As Ernenwein put it, there were political realities to consider: "Well, okay, how much can we push and how much do we have to accept? Several key players for the FAA used to work for the air-tour industry," he told me, "and a couple went back to the industry after a few years at FAA. The air-tour industry was also politically connected and donated some big money to Congress."

The draft noise plan presented for public discussion four alternatives to regulate overflights and chose one as the Park Service's "preferred option." Noise impacts would have been reduced if that alternative were adopted, Ernenwein said. But the reaction to the preferred option among most of the environmental groups ranged from melancholy to outrage. The industry's formal response accused the Park Service of trying to kill jobs and eviscerate air tours, the "death of a vibrant air tour industry by a thousand cuts."

At the time Park Service science and resource manager Jane Rodgers explained a parallel and far less public conversation that was underway. "I know of at least two attempts to bypass the public process through legislation at the congressional level," she told me then. For two years running McCain had introduced unsuccessful Senate legislation that would have established "natural quiet" in a new way: as the level of noise Grand Canyon already had. What could explain his pivot on this issue?

"It's hard as a public agency to watch that happen," Rodgers said. "You're trying to go through due process, and it is all about public engagement and public involvement. You have to do that. It's the right thing. It is incredibly stressful on the staff—people who are completely dedicated, committed, just integrity-all-the-way—to feel like their life's work is being put at the whim of a legislature. To have it all kind of just pulled out from under you.

"But the air tour industry has some significant lobbying power. They're doing what they do on their side, and we're forging ahead through the process, trying to be transparent." She didn't know it then, but at long last the noise issue would soon be resolved—abruptly so.

Within a few weeks McCain and the Nevada Democrat Senator Harry Reid, along with some other sponsors, offered an amendment to a high-way-funding bill. With an economy of language, their maneuver accomplished quite a lot. It short-circuited the Environmental Impact Statement process entirely. It took the power to decide the noise issue out of the hands of the Park Service. It decreed that the status quo at Grand Canyon is the desired state of affairs—that things are quiet enough as they are.

It altered the Park Service formula that specified that "at least 50 percent" of the park should be quiet, three-quarters of the time, by deleting "at least," effectively capping how much quiet the Park Service can push for in the future. And as explained earlier, only about half of the low level air traffic is even regulated, because of earlier exemptions.

The new law also included a plan of further concessions to the air-tour operators to induce them to adopt "quiet technology"—which will have a largely unknown impact on noise and affects only half of over-flights, in any case—by the year 2030.

McCain's written explanation for his work read, "Unfortunately, a process that began 25 years ago with the Grand Canyon Overflights Act of 1987 has resulted in a flawed Park Service Air Tour Plan." He added that the Park Service noise research was "faulty." He also echoed industry statements about threats to jobs and profits under the park's proposed plan. It was a humbling finale for the Park Service, which had offered a years-long series of compromise alternatives that in the end weren't deeply compromised enough.

Among the consolations offered by Senator McCain: "Air tours provide a unique sightseeing experience for people who might otherwise not be able to visit the Grand Canyon, particularly the elderly and the disabled." That compassion is also invoked by air-tour lobbyists. As it happens, "the elderly" do not make up a large portion of any air-tour operator's passenger lists at the Canyon, as the industry's own figures show, and only 1 to 2 percent of those passengers are disabled.

The South Rim offers a thirteen-mile-long, flat, paved, strollable, and wheelchair-accessible pathway with superb views of the Canyon. It is visited happily by hundreds of thousands of people in their seventies and eighties each year, and many with disabilities. But let's concede something, and gladly: it would be a very fine project for the National Park Service itself to provide helicopter sightseeing for the incapacitated, and it could repatriate the profits to Grand Canyon's ailing budget.

More realistic, though less poignant, data on what might actually be lost if aerial visits were abolished, or at least reined in, was provided by Alan Stephen, a spokesperson for the Papillon air-tour company. About 70 percent of helicopter-and-plane tourists are foreign visitors who don't have a lot of time to spend in one place, he told a reporter—they just want to get a quick bird's-eye view of the Canyon.

You can find quiet at some times and in some areas of Grand Canyon. But more than a quarter century after McCain's raft trip, you hear air-tour noise about 80 percent of the time in some areas, and up to forty-three separate "aircraft noise events" can occur in a twenty-minute interval. On the busiest days more than a hundred helicopters may be over the Canyon at once. Its reverberant rocks sometimes make a single aircraft sound like three or four, even if you can't see it, research has found, and aircraft noise can echo up to sixteen miles along the inner walls of the gorge. Not a single location in the entire million-acre park is completely free of aircraft noise all the time. That McCain-Reid–sponsored law, which certifies all this as "substantial restoration of natural quiet," is called "The Moving Ahead for Progress in the 21st Century Act."

CHAPTER THIRTEEN

Capture and Corruption

All Americans should be alarmed about the effects of money in politics. But it is conservatives who should be leading the fight for campaign-finance reform.

Richard Painter, professor, University of Minnesota
Law School

Natural systems on U.S. public lands are at once vulnerable and resilient—and so is the political system that governs their fate. Two intertwined, pervasive threats to both the public interest and public lands are now so familiar that we may even, *in extremis,* accept them as given: regulatory agency corruption and electoral corruption.

That's a large claim—especially the use of the spiky, charged word "corruption." We usually think of that as bribes or other illicit payment. But the corruption I refer to here, and which others explain in detail further on, is quite legal, and that's the problem. The national parks and all other public lands will ever be pursued by ardent claimants who think of them as available stalls in life's big marketplace. They are on the verge of succeeding in that historic, wholesale conversion now.

There are big hustles and small ones. One operator smuggled busloads of trail runners into Grand Canyon for years for a hefty fee but without a permit. They clogged campgrounds and trails—he told them to lie to rangers if asked whether they were part of a group—until he was caught.

On a different scale the then-superintendent Steve Martin decided to halt the plague of hundreds of thousands of throwaway plastic water bottles at Grand Canyon, following the example of successful programs at other parks. He made lengthy preparations to ban their sale and to substitute trailhead spigots with natural spring water to fill reusable bottles. They were installed for about $300,000.

But the Coca-Cola Corporation, bottler of Dasani and other labeled water, was discomfited. That was made known to the nonprofit National Park Foundation, to which Coke gives donations. Its president passed the message on to the director of the Park Service in Washington, the e-mail record shows, and he became concerned about the politics and the money. The bottle ban was called off, for further study.

Martin retired soon after. An environmental group wrested the pertinent e-mail traffic from the Park Service with a Freedom of Information Act request. Reporters tracked the story, and when they called, Martin detailed what had happened. Despite the telling e-mails, the Park Service denied that it had succumbed to improper influence, and Coke denied that it had peddled any. But the public uproar was easy to translate: Grand Canyon should not be for sale. You can't buy bottled water there now.

Unfortunately, you can indeed buy into or muscle into our political system in other ways that are decisive for public lands. Consider the symptoms we've already seen: the near-inevitability of more destructive waves of invasive alien species because Department of Agriculture regulators have to back away from effective import inspections; endangered species protection by the Fish and Wildlife Service and grazing regulation by the Bureau of Land Management that are submerged in politics despite clear-enough legal support for change; the acceleration of drilling and mining in ecologically fragile areas; ecosystems already unraveling due to climate shifts that go forward largely unhindered and unplanned for; and the near threat of wholesale conversion of federal public lands into state and private hands. We hear thousands of helicopters over the Canyon and swarms of snowmobiles at Yellowstone, where there should be undisturbed habitat for wildlife and solitude for people who seek it.

Overflight noise is our now-familiar example—the kind that undermines confidence in the priorities of elected officials and regulatory agencies. Recall the frustration of the air-tour industry in the late 1990s, when its lobbyist James Santini, himself a former congressman, called Sen. John McCain a "nemesis" on the noise issue. But Santini added, in a newsletter report for his clients, "We will continue to bang on the McCain door hoping someday to get inside his head with our rational appeals." As it happened, "someday" arrived rather quickly.

Because Santini also reported that, thanks to diligent industry attendance at McCain fund-raisers in Washington and Arizona, the senator had at least been listening. One source of cash was an iconic figure at the U.S. Air Tour Association: Elling Halvorson, the founder and owner of Papillon, the biggest air-tour operator at Grand Canyon and in the world. He also owns Grand Canyon Helicopters, Grand Canyon Scenic Airlines, Canyon Flight Trading Company, and several other companies. Halvorson "among other things . . . contributed funds numerous times to various Congressional fundraising efforts which helped USATA maintain much-needed access to key Members of Congress including Arizona Senator John McCain," the newsletter said.

And in a bill in 1998 to regulate air tours in all national parks, McCain took action that Santini praised as "dramatic, near miraculous." Among other things, the senator had seen to it that provisions of a new law that would have given the Park Service more authority in negotiations with the FAA were removed. Meanwhile, environmental lobbyists complained that their phone calls to McCain weren't being returned and appointments were canceled. Negotiations plodded forward for another decade.

In 2008 Senator McCain launched his second bid for the presidency, which quickly gained momentum in the polls and on campaign-finance ledgers. He now had allies in the air-tour industry. Elling Halvorson played a key role in McCain's campaign, as one of the high-level fund-raisers called "bundlers," who give more money to the candidates than anyone else. They arrange with family, friends, and associates to deliver checks to the candidate in one big bundle, and Halvorson's for McCain was in the range of $100,000 to $250,000, according to Federal Election Commission records. Halvorson and his wife contributed $85,000 directly to McCain's campaigns, most of it in 2008.

But there were Halvorson contributions to the Nevada Democrat Harry Reid's campaigns too. Senator Reid is a long-time air-tour supporter with a history of fighting flight and noise restrictions at Grand Canyon, including those that might impinge on tours originating in Las Vegas. One air-tour company engaged Reid's son-in-law to lobby on the issue.

We've seen the swerving path of McCain's decisions during the years of the Grand Canyon noise controversy and the record of his gift taking from air-tour interests plainly enough. So tell me what you think I insinuate here: that in these dealings, McCain has proven himself a senator of easy virtue? I don't.

McCain's office declined my request for an interview, as did the helicopter mogul Halvorson. But I could not have judged the degree to which McCain's actions have been influenced by those or other major funders, no matter what he might have told me. We can't x-ray his heart and mind to sort out his motives. In any case, McCain has a long record of indignant support for campaign-finance reform. It is not that he is personally corrupt, certainly not under current law. Instead, his public service unfolds within a corrupt system, one that has made most of us skeptical, if not cynical.

"And until we create the conditions under which trust is possible—when, in other words, the presence of money in the wrong places doesn't inevitably make us doubt—this skepticism will remain," the Harvard law professor Lawrence Lessig writes. We would all be better off, the senators included, with a firm set of rules to fix that problem.

Meanwhile, we think we smell something bad, and the fragrance is so pervasive that an emphatic majority of citizens have lost faith in the legislative process. In a recent poll, 75 percent of Americans said they believe "campaign contributions buy results in Congress." That's a margin of three to one, and Republicans (71 percent) were nearly as alienated as Democrats (81 percent). "Puzzles plus money produce the view that the money explains the puzzles. In a line: We don't trust our government," Lessig, who commissioned the poll, has written. Law professor William Black, a former bank regulator, summarizes the ordinary citizen's street-level, tragic view when he writes that a campaign contribution always generates the best return on investment.

Real estate developers and the purveyors of thrill rides like snow-mobiles and helicopters will always be poised to take advantage of revenue opportunities offered by national parks, just as timber harvesters, miners, and gas drillers will on other public lands. Maybe the citizenry decides that's okay in some cases—or, at other times, those interest groups may be stymied by public opposition. But generally, we're too busy with the rest of life to monitor the national interest in detail. That's why we expect—or used to expect—to elect focused, conflict-free legislators who would represent us.

The political scientist Dennis Thompson points out that merely because we have accustomed ourselves in the law and in our political life to large-scale campaign donations does not mean that they are less potent, nor less corrupt, than the sort of bribe that pays a public official in cash for a vote or some other act: "Corruption that works through patterns of conduct, institutional routines, and informal norms may leave fewer footprints, but more wreckage in its path." He adds that "citizens have a right to insist, as the price of trust in a democracy, that officials not give reason to doubt their trustworthiness."

Fordham University law professor Zephyr Teachout writes that for public officials "a gift can be a bribe. A bribe can be a gift.... They can create obligations to private parties that shape judgment and outcomes." She has traced the halting development of our national outlook on what is and is not corrupt back to a time just after the American Revolution. It was a central preoccupation of the new republic then, as it would often prove to be in later years.

For example, an embarrassed Benjamin Franklin decided to relinquish to Congress a diamond-studded snuff box, a gift from the king of France, after an outcry about how poorly it reflected on the integrity of his public service as ambassador. It wasn't just the possibility that Louis XVI might have deflected Franklin's focus on the United States' best interests. How could we know that, one way or the other?

The problem instead was the appearance of the act, which is, in other than exceptional cases, all the public can go by, in judging the honesty and integrity of an officeholder. Accepting gifts like huge campaign donations opens the possibility of what we commonly think of as a conflict of interest, of divided loyalty, of trying to serve two masters. And the appearance of conflict alone compromises our trust. It has in the past and it does now, on an epic scale, in the carnival of conflicts that is the Trump White House.

When venality in the U.S. government is deep and frequent, some of our leaders begin to resemble, say, Uzbekistan's. Our cynicism may enlarge, but it shouldn't. We've been here before. Reform campaigns of a century ago cleaned up the bullying of civil servants, nepotism, bribery, collusion, conflicts of interest, and many other forms of public corruption. But it's not as if that battle ever ends. That is why public service, and so many kinds of private employment, have rules against taking gifts from interested parties. In Franklin's time and until fairly recently, for example, campaign donations were considered to be close kin to bribes, and to accept them was to risk being seen as corrupt.

We have always known this. Mississippi Democrat John Stennis was, during the 1970s, chair of the Senate Armed Services Committee, which oversaw the spending of hundreds of billions of defense dollars. He was asked to hold a fund-raiser at which defense contractors would be present, and he refused. "Would that be proper?" he asked. "I hold life and death over those companies. I don't think it would be proper for me to take money from them."

Barry Goldwater, Arizona's senator for thirty years and a Republican presidential candidate, also clarified what once was obvious: "Senators and representatives, faced incessantly with the need to raise ever more funds ... can scarcely avoid weighing every decision against the question, 'How will this affect my fund-raising?' rather than 'How will this affect the national interest?'" If nothing else, politicians who grovel for special-interest money tend to disgust the public, he said.

And then there is Senator McCain. "Questions of honor are raised as much by appearances as by reality in politics, and because they incite public distrust, they need to be addressed no less directly than we would address evidence of expressly illegal corruption," he wrote in his 2002 memoir *Worth the Fighting For.* "By the time I became a leading advocate of campaign finance reform, I had come to appreciate that the public's suspicions were not always mistaken. Money does buy access in Washington, and access increases influence that often results in benefiting the few at the expense of the many."

Gifts have great power, even when they seem more trivial than Franklin's snuff box. Several studies have demonstrated that even freebies like ballpoint pens, lunches, and ad-bearing pads of sticky notes from pharmaceutical companies can influence doctors' choices of which drugs to prescribe, and how often.

A handful of scientists who have taken money from oil, coal, or electric utility companies routinely produce research and public commentary on global warming that duplicates the views of their funders. We put less store by their work when we know the sources of their income, which explains why those sources have often been carefully hidden from public view. Most scientists try to avoid such conflicts of interest.

Sometimes the law forbids conflicted behavior in important settings. Doctors cannot refer Medicare patients to a heart-scan clinic or a laboratory services corporation in which they have a financial interest. Reputable news media have codes of ethics for journalists—both reporters and opinion writers—that anticipate the issue by forbidding gifts from sources and any outside business relationships. Violating the code can be a firing offense.

This is elementary logic, no matter how neglected in our political contests. If your architect takes a vacation cruise paid for by a window manufacturer, even if she swears she is absolutely trustworthy, you are troubled if not outraged. If a Supreme Court justice goes duck hunting with a litigant in a Supreme Court case, or if state legislators take junkets to France that are paid for by a uranium-mining company whose project they'll vote on, more than mild suspicions are aroused that they may be playing crooked games. Those examples—real and recent—deal hard blows to public confidence.

Teachout likens this to the "rotten boroughs" of England in the 1700s, when the few voters in a largely abandoned place like the village of Old Sarum were accorded as much representation in Parliament as a major city like Manchester. No one had updated the boundaries of voting districts to reflect where people really lived. Holding all that leverage, the few voters in these places often sold their votes. Money displaced individual citizens as the source of legislative power.

In our current system a tiny number of voters are also disproportionately represented in Congress and in state legislatures, because their voting strength is multiplied by their campaign-donation dollars. The money buys them astronomically more weight in the scales of power than your vote. The Tusayan airport at Grand Canyon, with its big fleet of brightly painted helicopters, has become Old Sarum—a rotten borough where the source of dollars can outvote the public.

In public service as in medicine, journalism, or architecture, the best cure for conflicts of interest is full disclosure and an outright ban, relentlessly and transparently enforced. Here's my own disclosure, for you to judge: I have contributed money to Democrats and environmental groups, on occasion. I have never taken a dime from any of them. Money awards and speaking fees are returned or donated.

We are no longer on the Ben Franklin program in terms of campaign donations to candidates for federal public office in the United States. We do not consider those gifts as bribery. We should. Our national legislators are spending far too much time—as much as 30 to 70 percent of it—soliciting campaign money instead of using that time to research, contemplate, and pursue the public interest.

As lobbyists like the U.S. Air Tour Association, and many legislators too, have readily if not gleefully acknowledged, those gifts buy "access" for the givers—access that you and I and other nongivers are not nearly as readily granted—if indeed we can gain any access at all. While it cannot be demonstrated that certain legislators cast a vote only because they received a campaign donation, it beggars belief to think the odds may not be altered, over time, as a destructive dependency flourishes.

Lessig and Teachout, for example, point out that fewer than 2 percent of U.S. voters gave anything to any political campaign in 2014. The top hundred donors gave as much as the bottom 4.75 million. That's a system that produces a special-interest-dominated and dependent—a corrupt—Congress. The authors advocate new federal legislation that could quickly bring reform: overhauling lobbying and ethics laws, creating strong incentives for small donations—"citizen-funded" elections—and dramatically increasing transparency.

"It's the way things work now," as we frequently hear about these matters, and it gives frontline public servants cause to doubt the value of their own efforts and the honesty of the whole governmental system. At Grand Canyon, our prime example, at least mild regulation of noise might have been imposed. When Congress passed the legislation that sidestepped that whole public process, Senator McCain dismissed the Park Service research out of hand, saying it "relied on faulty assumptions." The former Grand Canyon senior planner Rick Ernenwein figured that McCain "basically took something the air-tour industry had put together as their major comment and just signed his name to it."

McCain's pivot toward the air-tour industry, Ernenwein figured, coincided with his run for president: "I think he developed a lot of IOUs." As for the final act, "It was an eleventh-and-a-halfth-hour thing they slipped in, that nobody was able to see in some must-pass legislation. So yeah, there was a lot of behind-the-scenes politicking that went into that."

The superintendent at the time, David Uberuaga, had to figure out what the park's next move could be, and on the noise issue he was nearly out of moves. "We spent millions of dollars on soundscape management to deal with the air-tour industry, so we could say, here's what 'substantial restoration of natural quiet' over Grand Canyon is,'" he told me. "And then we get one piece of legislation with one paragraph, because the lobbyists were able to influence that outcome. They shot all that work right out of the water. We had enough science, but I wasn't able to effect the change that I needed."

We met the noise researchers Laura Levy and Sarah Falzarano in the previous chapter. Their years of sweltering fieldwork, meticulous record keeping, and celebrations of owls, coyotes, and slobbering elks was part of that science. They, too, took note of the final result of their public service.

Levy left before Grand Canyon's proposed regulations were completed. But by chance she had returned there as a volunteer for a raft trip with a group of teenagers when she heard the news that the whole process had been ditched in Congress. "I was very happy at that point that I had decided to go to grad school and not continue working on something that was ultimately settled with a handshake—a deal made behind closed doors. You know about that kind of stuff, but at the same time it's really sad when politics overtakes science. That's the take-home message for me," she said. Her colleague Falzarano recalled, "I tried to keep my head down and do good science. But it definitely was sad that everything we did was set aside. That was a big part of my life and my professional work."

Their supervisor, Ken McMullen, told me that by that time, "I realized that good science didn't matter. Politics mattered more. By then I was pretty much, you could say, demoralized, I guess." As for outside political influence, and endless foot dragging by the Federal Aviation Administration, "We all knew what was going on. That was one of the things that convinced me to get the heck out of there and retire. It went from good science to political expediency." For years the Park Service had been forced to give ground, he said, but it led only to capitulation by Congress, first to the Federal Aviation Administration and then to the air-tour industry.

A series of air crashes, including a midair collision of two airliners over Grand Canyon, led to the creation of the Federal Aviation Agency—later, the Federal Aviation Administration—in 1958. The founding legislation gave the FAA a "dual mandate": responsibility for promoting commercial aviation as well as ensuring its safety.

That deeply conflicted role was in force when the original Grand Canyon overflight noise and safety legislation was passed in 1987. And during the subsequent years of negotiation over noise levels at national parks, the FAA was often seen as a firm ally of the air-tour industry by the Park Service personnel I've spoken with.

If true, that would be a classic case of "regulatory capture," which legal scholars have defined as a "form of political corruption that occurs when a regulatory agency, created to act in the public interest, instead advances the commercial or political concerns of special interest groups."

An FAA spokesperson contradicted that view of its role at Grand Canyon, however. Overflight regulation at the park "includes limitations on the annual number of air tours that are allowed to be conducted, mandatory routes for air-tour aircraft, curfews, flight free zones, and general aviation corridors," I was told. "The FAA did not represent or speak for air-tour, airline, or general aviation representatives that participated in this process; they represented themselves." The agency also points to a lengthening record of comparatively crash-free aviation in the United States as potent evidence of its effectiveness as a safety watchdog.

It's useful, then, to trace what was happening at the FAA during the long years from the mid-1980s to 2012, when it says it acted as a neutral safety enforcer during negotiations with the Park Service, the air-tour industry, and other stakeholders. By 1996 a series of air crashes rocked the FAA and led to federal legislation that eliminated "promotion" from the agency's job description.

Or seemed to. The change was actually just cosmetic, critics charged, and an addendum to the new law seemed to confirm that view: "the managers do not intend for enactment of this provision to require any changes in the FAA's current organization or functions."

And a dozen years later, in 2008, the FAA's deeply ingrained reflexes favoring promotion for commercial aviation had not changed, according to testimony before the House Committee on Transportation and Infrastructure. FAA safety inspectors said they had been silenced and threatened by superiors when they tried to alert the agency to serious safety issues. Others testified that a "revolving door" led from regulatory jobs at the agency to more lucrative employment at the airlines.

The FAA mission statement on its website at the time included "being responsive to our customers and accountable to the public." The "customers" were the airlines—and that seemed to many to summarize the agency's priorities all too clearly. The airlines "constantly remind us they are the customer," one inspector testified. "The best way to put this is like you are going down the highway committing traffic violations and jeopardizing the safety of others, and when the police officer stops you and informs you that you are breaking the law by endangering people's lives, you tell the officer that he cannot document the violation because you are his customer."

The inspector said he had alerted superiors about fuselage cracks in one airline's fleet that might have proved "catastrophic" in flight, and his findings were suppressed. The aircraft continued to fly for more than a year, he testified. That testimony was confirmed at the hearing by the inspector general of the Department of Transportation.

"I am here to report that more than one FAA inspector, along with FAA management, have been looking the other way for years," one whistleblower, Charalambe "Bobby" Boutris, told the committee. "No supervisor can do what my supervisor was doing without the support from fellow inspectors, the support of the division management team, who were fully aware of what was going on, and I believe with the support from some people in Washington."

A second whistleblower testified that when he raised safety issues, his job as a longtime inspector and that of his wife, in another part of the agency, were threatened. He said he had submitted a fifteen-page written statement describing more than two years of improprieties, unethical actions, abuse of authority, misuse of government resources, and relaxed oversight. Months passed, and "it was evident that no action was being considered," he testified.

The congressional committee chair assailed the "culture of coziness" among senior agency officials and the airlines. "FAA needs to clean house, from the top down," he said. "The FAA would have us believe that what took place was an isolated incident and has been contained. In fact, the testimony we have heard substantiates that, clearly, this is not an isolated aberration attributed or attributable to a rogue individual, but, rather, a systematic breakdown of the safety oversight role of the FAA. It is misfeasance, malfeasance, bordering on corruption. If this were a grand jury proceeding, I think it would result in an indictment."

I asked the FAA for a detailed response to the hearings, and its spokespersons declined to provide one. After the hearings the agency's next exercise in cosmetology followed: its mission statement was amended to clarify that its "customers" were no longer the airlines.

A couple of years after those hearings, the draft plan for overflight noise regulation at Grand Canyon was released. The FAA, meanwhile, was in more trouble, this time about its lack of regulation of regional air carriers, whose planes were crashing far more frequently than those of the major airlines. Mary Schiavo is a former inspector general of the U.S. Department of Transportation and is now a plaintiff's attorney. She told a reporter, "They might be good cheerleaders, but they're terrible cops.... The FAA protects airlines."

After Congress had scuttled the Canyon noise plan and the public process that generated it, Grand Canyon's superintendent did not come away with the feeling that the FAA had been a neutral arbiter. "The FAA and industry have put their energy here, and they've busted their butt every chance they can, because they think whatever we get will be precedent-setting for all national parks. So that's how they focus their resources," he told me. "It didn't matter what the science said. We built that up, and we were ready to go public, and they came with an end run politically that took the science out of the equation."

Beholden members of Congress, who may be attuned far more alertly to lobbyists and campaign contributors than to the public good, enable this other kind of corruption of public agencies: regulatory capture, the obeisance to political clout. We've encountered it at the Bureau of Land Management, the Agricultural Plant Health Inspection Service, the U.S. Fish and Wildlife Service, the Federal Aviation Administration, and the Park Service itself at times. After all, agency administrators who impose burdens on the industries they are supposed to regulate invite intervention by congressional representatives.

Just as our current system of campaign financing breeds cynicism, so does regulatory capture. It dismays the center, the left, and the right of our political discourse, the editors of a recent book have written: "As recent polling suggests, this fatalism has over time deeply influenced not just scholars and inside-the-Beltway cynics, but the broad mass of the American electorate."

Public trust in the federal government has declined steadily over the past forty years and is now under 20 percent. Liberals see regulatory capture as a recurring, widespread, serious problem that governmental reforms can mitigate. Regulation addresses otherwise insoluble problems, so it has to be fixed, they conclude.

Conservatives share the concern but often argue that capture points toward what is, for them, an obvious solution: less regulation. "All Americans should be alarmed about the effects of money in politics. But it is conservatives who should be leading the fight for campaign-finance reform," Richard Painter of the University of Minnesota Law School has written. "Why should conservative voters care? First, big money in politics encourages big government.... When politicians are dependent on campaign money from contractors and lobbyists, they're incapable of holding spending programs to account." Campaign contributions also breed more regulation, Painter argues. Companies in heavily regulated industries such as banking, health care, and energy are among the largest contributors to push for regulations that freeze out competition.

Whether your vote is for effective regulation or less regulation, a critical question is whether capture can be avoided. "The widespread belief that special interests capture regulation, and that neither the government nor the public can prevent this, understandably weakens public trust in government and contributes to a sense that our political system is not capable of meeting the challenges it faces," as political scientist Daniel Carpenter and economist David Moss summarize in their anthology, *Preventing Regulatory Capture*.

Can government agencies elude capture? "We believe the evidence strongly suggests that the answer is yes," these authors have concluded, and they recommend a range of strategies that have proven to be therapeutic in the past: "Regulatory capture is not always and everywhere the devastating problem it is often made out to be," they write. "In some cases, good regulation does prevail, in spite of the special interests.... Many regulatory bodies have developed surprisingly strong immune systems, apparently capable of keeping the worst forms of capture at bay."

We've seen in prior chapters that the U.S. Fish and Wildlife Service, the Bureau of Land Management, and other agencies sporadically respond to litigation, or the threat of it, with better management of endangered species or other natural resources. The judiciary is usually more insulated from political bullying and grooming, but it is not a panacea. There can also be useful oversight by journalists, the authors add. But that's an increasingly faint hope in these days of financially enfeebled news media, their shrunken ambitions, and the expense of investigative reporting.

Here, we've dug down to bedrock. It is difficult to imagine that mere tinkering will reanimate rusting mechanisms of governance. Without muscular, enduring public support for major reform, these agencies, the media, the legislatures, and even the courts are increasingly forfeit. An undertow pulls at hope when good government is continually betrayed. That feeling is especially widespread now. But reform has played a larger role in our politics than decay, over time. An interested public is the strongest force to make agencies pursue the public interest: revolving doors can be locked; political-insider pressure can be met with public pressure; the role of money, muted.

So maybe under the press of distraction and distress we'll give up on all that, but I doubt it. Our national history belies terminal hand-wringing. That's how the Progressive movement won the political system back from the robber barons a century ago. Reclaiming public resources also recurs, forcefully. That's how our system of national parks, forests, monuments, and wilderness areas, the Endangered Species Act, and the Clean Air and Clean Water acts—all that we've inherited—came to be.

CHAPTER FOURTEEN

The Lost Orphan's Legacy

We've done it for dams and reservoirs and highways, why not for natural areas?

Reed Noss, conservation biologist

The New Hance rim-to-river trail isn't new. It's about 125 years old, unmaintained, rough, and steep. The trailhead is not a typical entry path either. It's more like one of those infinity pools: you see the Canyon edge and then just empty space. I stepped off into a chilly dawn with a strong young Uzbek friend, Sanjar, who was celebrating his new college diploma, and Rohan Roy. He's the kind of hiking guide who, with his wife, in a single two-week, nonstop, late-winter backpacking trek, recently ascended and skied a string of fourteen Colorado peaks over fourteen thousand feet high.

On one high skinny section, the New Hance trail bends in a tight horseshoe, mobile rock fragments underfoot. We brushed past a rough sandstone wall on the right, past a sheer, dizzying drop on the left. We saw, up ahead, oh, maybe a four-and-a-half foot gap where the trail had fallen away. I had heard of this spot. A year or so earlier a hiker had been lucky enough to falter here, miss her hop-stride, tumble through for a long fall, shatter an ankle, and be evacuated out by helicopter. She might easily have died. As we shed our packs to manage this transit I caught Rohan's eye and thought, "Okay, we've followed along this far. Are you kidding? *Now* what do we do?"

All went well. But if you've read this far, the same question may have occurred to you. This book is littered with prescriptions for what to do about wildlife and global warming and, before we forget, a sorely afflicted political system. If I type, "How can I save public lands?" in a browser window, I get 40,100,000 suggestions in 0.82 seconds, and the first couple of screenfuls are pretty good, too. But the overall question hasn't really been answered, and can't be in any general sense, so I'm ducking it entirely. It's a more focused, individual thing: what's the very most effective way for a particular citizen to leverage energy, wit, indignation, patience, money, and time? You can figure that one out.

A few hours down the trail, halfway between the big conifers on the high saddles and the gravel washes near the bottom, we found some shade. From this rocky promontory with a view out over Red Canyon, we noticed a scatter of black chunks on the ground nearby—a tarry conglomerate called breccia, a common signal of uranium deposits in the region.

Grand Canyon National Park had its own uranium mine awhile back, the Lost Orphan, not far from Grand Canyon Village. Some of its ore was among the purest ever found in North America. The mine ceased operation in 1969, but its headframe stood for nearly forty years, too radioactive to leave standing but difficult to safely take down. Finally it was removed, along with many tons of contaminated rock and soil. Horn Creek, along the Tonto Trail below the mine, will be too "hot" to sip from safely for thousands of years. But the Lost Orphan left a positive legacy too.

When the price of uranium jumped a few seasons back, thousands of mining claims were staked on federal lands south and north of the Canyon. Studies were launched to try to figure out whether a wave of new mines might contaminate underground water sources. The city of Los Angeles, worried about the Colorado—much of its water supply—posted formal objections to new uranium mining with the federal government, along with several tens of thousands of others around the world. Though it was by no means the only compelling argument in this discussion, the example of the Lost Orphan was frequently cited: radioactivity is tenacious, and water, especially, carries it along. Lesson learned.

Three lessons, actually. A surge of concern for the park resulted in the withdrawal, by the secretary of the interior, of about 1,560 square miles of federal land from new mining ventures until the year 2032, when the decision will be revisited. Public opinion matters. It did back in the 1960s, when the Lost Orphan's owners circulated a proposal for a six-hundred-room, twenty-two-story hotel, to be pasted onto the flank of the Canyon right under its mine operation. National outrage scuttled the idea.

We've seen that happen in these pages too, when the Disney corporation backed away from its theme park amid protected Civil War battlefield sites in Virginia. It mattered during the too-recent Paul Hoffman era, when a parade of destructive changes was proposed for national park operations. It did when the Grand Canyon Escalade proposal seemed to melt in the glare of negative public opinion.

A last take-home, however: no decision is final. Public lands will always be the targets of pressure from private interests and their political comrades-in-arms. The National Mining Association and several coplaintiffs are suing in federal court to have the uranium mining ban lifted. Washington has now returned us to the philosophy that the government should "put our public lands back to work." The Trump administration, which has favored more mining and drilling on federal public lands, may well have reversed course to allow uranium mining to start up in the Canyon region again by the time you read this.

The future of the town of Tusayan and its potential impact on Grand Canyon's water sources offers the same contingencies. Tusayan got an answer to its request for an easement across Kaibab National Forest land, which was needed for the proposed real estate development we looked over in an earlier chapter—a thirsty Gargantua near the Canyon rim. The answer was no.

During its deliberations, the Forest Service received more than 37,000 unique comment letters, 85,693 form letters, and two petitions with 105,698 signatures. The overwhelming majority opposed the development, the Forest Service district supervisor's letter told the town council. "Based on information received in the record, I have determined that the Tusayan proposal is deeply controversial, is opposed by local and national communities, would stress local and Grand Canyon National Park infrastructure, and have untold impacts to the surrounding Tribal and National Park lands," she explained.

Once again, however, there is never a Last Word. I asked Michael Dombeck, former chief of the Forest Service and former director of the Bureau of Land Management, to put himself in the shoes of Tusayan's Italian real estate developer, Gruppo Stilo, and look for a path around this rebuff. "The options are that they could appeal the decision to the regional forester, and it could even go to the chief's office in Washington," he told me. "They can also go to court. The third thing is that they can politicize the issue. They can draw their congressional delegation in and begin to put pressure on the administration to push the decision their way, if they can." Donald Trump, he remarked, is a real estate developer.

Dombeck added, "This is why it's important for as many citizens to get involved as possible. If you have wealthy developers who can hire the best attorneys and the best lobbyists in the land, they can run circles around a few citizens who don't have the resources to be competitive or even play in this game."

Shortly after the Forest Service's decision on the Tusayan megadevelopment, former park superintendent David Uberuaga told me, the Italian embassy's economic team in Washington had contacted the Park Service and the Forest Service on behalf of their compatriot real estate developer, Gruppo Stilo. They asked for some kind of planning session. They were advised to approach the Department of the Interior instead. "All the developer wants," he told me, "is to be able to say, come to the Grand Canyon. I have Grand Canyon in my portfolio, and I'm cool. It's just a recess for them."

Hours later we reached our campsite at the Hance Rapids. We shed our dusty packs to snore, in my case, among the tamarisks along the sandy river shoreline. The temperature had hit 106. No question whether we were going to plow into the Colorado to cool off, and that cooling was instantaneous when we did. The river here is a brisk 46 degrees, cold enough to kill you in not very many minutes. It emerges upstream from the big pipes near the bottom of Glen Canyon Dam and the frigid depths trapped behind it, in Lake Powell.

Before the dam was completed in 1963, the Colorado was around 80 degrees at this time of year. Since then those newly cold flows have helped drive out three of the eight native fish species that used to occur here. Three others are on the endangered list: the magnificent Colorado pikeminnow, which could grow to a hundred pounds and six feet in length; the humpback chub; and the razorback sucker. Another major factor in their decline is the presence of destructive invasives. Some of them, like brown and rainbow trout, were introduced on purpose as game fish.

Glen Canyon Dam blocks the natural flow of sand along the river too. Some of the bottom has been scoured to bedrock. Loss of beach sand has exposed archaeological sites that once were protected. Downstream, the absence of natural flows has resulted in heavier tamarisk invasion and disrupted and depleted wildlife habitat.

A sober case has been made that the dam should be breached to loose the Colorado. On more than a thousand U.S. rivers that has occurred, with excellent results for fish migrations and the rest of their natural systems. The dam does generate electricity, for now. It also creates Lake Powell for houseboaters to punt around on. But some hundreds of billions of gallons of increasingly precious water are lost each year to evaporation on the hot surface of Lake Powell and from absorption into the surrounding sandstone. Finally, the lake has completely submerged Glen Canyon, said to have been as beautiful as Grand Canyon before it vanished.

Colorado River water has been parceled out to an unrealistically long and thirsty list of claimants, and they all vie to maintain or expand their legal shares: California versus Arizona, farmers versus urbanites versus Native American tribes, Upper Colorado Basin states versus

Lower Basin states versus Mexico. But the current drought in the Southwest has already diminished flows and drained water levels at Lake Powell so low that there's a small chance they could be below the power generating turbine intakes as early as 2018.

Downstream at Lake Mead, near Las Vegas, two Scripps Institution researchers have projected that water commitments to too many users and the impacts of climate change combine to yield a 50 percent chance of an even more drastic event by the year 2021: "dead pool," when power generation ceases and the river flow itself is threatened. One of that study's authors, Tim Barnett, told me that letting the Colorado dry up below the dams is an extremely remote possibility, however. "The engineers are not going to let those lakes go to dead pool. They'll stop enough withdrawals to keep the river flowing. They'll keep cutting back," he said. Nonetheless, water levels anywhere near that vicinity are a haunting scenario. "I have this picture in my mind of the last days of the Colorado, and there's a mere trickle," Barnett said. "Battalions of lawyers are down in the middle of the muck throwing mud at each other and rolling around and basically arguing over a dead resource."

Now's the time for all those Colorado users to hash out an agreement instead, and the opportunity is slipping away, he said. "Today, we are at or beyond the sustainable limit of the Colorado system," Barnett and his coauthor, David Pierce, have written. "The alternative to reasoned solutions to this coming water crisis is a major societal and economic disruption in the desert southwest; something that will affect each of us living in the region," their report concluded.

Not long ago decommissioning Glen Canyon Dam to free the river through Grand Canyon was reflexively dismissed as a radical environmentalist's pipe dream. It was not on the list of options for the future drawn up by the National Park Service and the U.S. Bureau of Reclamation, which operates the dams. Part of the mantra: we need the electric power generated by those big turbines to keep costs down for about three million regional users.

On this front too, however, climate change is starting to bulldoze the politics of the status quo. I asked at the Department of Energy's Western Area Power Administration Headquarters for an estimate of how much more electrical energy consumers really could expect to pay if the dam went away. Never got a straight answer. Not even close.

The Glen Canyon Institute commissioned a group of economists to answer that question and a few others, though. It has long campaigned to draw down Lake Powell and use the dam only if Lake Mead, downstream, fills up (which next to no one is predicting). The institute-sponsored study concluded that the dam's electric energy represents less than one half of 1 percent of the sales value from electric generation in the power grid that serves the western United States. Average monthly cost increases would be a trifling eight cents for residential customers, 59 cents for commercial customers, and $6.16 a month for industrial consumers of Glen Canyon Dam electricity. Shutting down the dam could save nearly $75 million a year in management costs, potential increased hydropower downstream, and conservation of water that would have otherwise have been lost through seepage behind Glen Canyon Dam.

Dan Beard, commissioner of the Bureau of Reclamation during the Clinton administration, goes further. He proposes not only that Glen Canyon and other "deadbeat dams" should be torn down, but that the bureau itself should be abolished. On behalf of the "water nobility" that profits from this mess, the agency and its portfolio are environmentally destructive, utterly unnecessary for power generation, and wasteful of precious water and of tens of millions of dollars each year, he has written.

So the future of Glen Canyon and the Colorado River joins those other challenges: invasives, endangereds, air and water pollution, grazing, fossil fuel extraction ... and, above all, climate change that may leave some of our most prized national landscapes as biologically barren as Haiti.

A blue-ribbon Park Service advisory committee recommends making the parks the core of a national public-lands conservation plan. "Connectivity across these broader land- and seascapes is essential for system resilience over time" to support animal and plant migrations and to prevent destructive inbreeding due to isolation, the report said. It recommends the amalgamation of larger public landscapes and longer time horizons for managing them.

Private lands, too, already play an essential role in the survival of endangered species. The role will expand as the climate becomes hotter. Reed Noss, a conservation biologist at the University of Central Florida, proposes, "In general I don't like to go on record as a proponent for condemnation of land as a blanket solution, but I think it has to be part of the solution. We've done it for dams and reservoirs and highways, why not for natural areas? The government has done virtually nothing on land acquisition. Everything has its price, and we could just buy land." For what has already been spent on the Iraq War, a million square miles of land could be purchased and safeguarded for its environmental value to humans and to wildlife, he calculates.

Gary Machlis, science adviser to the director of the national park system and one of the authors of that advisory committee report, told me that conservation should be the top priority "use" for all our "multiple use" resource agencies. That includes, of course, the national park system, which often exhibits other priorities. The BLM and the Forest Service and the Fish and Wildlife Service embrace a "stakeholder" operational mode now. Its conjecture is that things work out okay when all stakeholders use their influence to slug it out. That way of looking at priorities may protect the agencies' interests, but it does not protect the resources. "I think all the agencies could have as the core mission the preservation of functioning ecosystems," Machlis says. "And we'd all be better off. It would drive us toward a coherent management policy in the United States."

That would mean giving up the political pretense that we can have them all: conservation, industrial extractions, grazing, roading, and shooting. We can't protect the nature and the future of public lands if we continue selling them off to those influencers too. Almost too obvious to mention, though, is that science does not always translate into power. Research has already registered abundant and ominous warnings. That doesn't by itself make natural systems the highest priority for public lands.

Instead, the hopes that remain are a prize won in 130 years of conservation battles, led by both Republicans and Democrats at times. Dispiriting, or perhaps inspiriting: if we want that waning legacy to endure, we're going to have fight hard for it. As Martha Hahn, former chief of science administration and resource management at Grand Canyon, says, "What is needed is for the public voice to rise to a level even deaf ears can hear."

NOTES

NOTES FOR ILLUSTRATIONS

Map 1. Public lands in the United States: The map is from U.S. Geological Survey, *National Map.* The reported land area figures, which may differ from what is depicted on the map, are from U.S. Forest Service, *Land Areas,* 1; U.S. Fish and Wildlife Service, "Statistical Data Tables"; National Park Service, *Land Resources Division Summary; Vincent, et al., Federal Land Ownership.*

Map 2. Some places described in this book: U.S. Geological Survey, *National Map.*

Map 3. Shifting climate, shifting nature: These maps were created by climate change–applications developer Chris Zganjar of The Nature Conservancy. He applied an 8.5 Representative Concentration Pathway—basically the "business as usual" scenario, in which we continue to accelerate greenhouse-gas concentrations, worldwide, at current rates. He generated these distances using a median of the twenty-six climate models and the Climate Wizard Custom Analysis Tool.

Map 4. How much hotter will national parks be? These maps were created by the climate scientist Katharine Hayhoe, director of the Climate Science Center at Texas Tech University, and geospatial scientist Sharmistha Swain, also of Texas Tech. The maps are based on those in Stoner et al., "Asynchronous Regional Regression Model."

Map 5. Western public lands used for grazing: Map by Kurt Menke, Bird's Eye View GIS.

Map 6. The health of the Bureau of Land Management's grazed land: This map is based on BLM data gathered by Peter Lattin, a former BLM geographic information–systems contractor and project leader for BLM Rapid Ecoregional Assessments. It provides only a broad general picture. Lattin emphasizes that because BLM assessments are frequently in flux, the map and associated estimates of land health change frequently. Land health status given here represents the most recent evaluations (2012) of sampled acreage the agency has released. BLM allotments shown include some private and state-owned land as well as federal public land.

The agency's data are incomplete and inaccurate to an unknown extent. "I made no attempt to clean up BLM's mess and release[d] these data 'warts and all' to help bring some transparency to the agency's data quality issues," Lattin told me. "Accuracy and precision is a problem, but the big picture is not materially diminished.... The user should take these limitations into consideration."

Lattin says the agency has included these same data in a summary of the risk livestock represent to an endangered species, the greater sage-grouse, and in one of its recent Rapid Ecological Assessments.

"Based upon my examination of ground conditions using high resolution aerial imagery, many locations across the West and identified as 'meeting all land health standards' have disturbance levels that suggest that they could not meet land health standards," Lattin adds. "What fraction remains to be determined."

The BLM's response is that acreage sampled in an allotment is not the same thing as the whole allotment. So mapping a whole allotment as meeting or not meeting health standards may be inaccurate if the sample is not representative. Richard Mayberry, of the Division of Forest, Rangeland, Riparian and Plant Conservation, adds, "because the Bureau has not created its own map for comparison purposes, we cannot determine accuracy of the map."

Until the agency completes its own map of which of its lands are in good shape and which are not, then this map is state of the art.

Map 7. Linking public lands for survival: The proposed network actually extends from Mexico into Canada. Map by Ron Sutherland, Wildlands Network.

Map 8. Proposed Grand Canyon Heritage National Monument: Map by Kim Crumbo, Grand Canyon Wildlands Council.

NOTES FOR CHAPTERS

Chapter One. Brink

Two billion years of geology: National Park Service, "Geology Field Notes."
Most severe drought the region has seen: Jeffrey Dean, interview.
Overgrazing and eroding fragile rangeland: Martha Hahn, interview.

28 percent of the national dirt: Vincent et al., *Federal Land Ownership,* i.

20 percent of the United States' clean water: U.S Forest Service, "By the Numbers."

The U.S. House of Representatives voted: H.R. Res. 5. "Treatment of Conveyances."

Nothing could be more American: Busse, "House Places No Value."

Why do they [the feds] own all that land: Hannity, "Ranch Standoff."

A shit-ton of land: Pearl, "Armed Militiamen."

What you can do: Roosevelt, *Compilation of the Messages,* 327.

To conserve the scenery: Dilsaver, *America's National Park System,* ch. 1.

America's best idea: MacEachern, "America's Best Idea."

The secret root: Holmes, *Common Law,* 36–37.

The jumblestack of habitats: National Park Service, "Grand Canyon Ecosystems"; Anderson, Hirt, and Gerke, "Nature, Culture and History."

Formally recognized as extinct: National Park Service, "Sensitive Wildlife."

Chapter Two. Alien Abductions

10 line-of-sight miles: National Park Service, "Backcountry Trail Distances"; National Park Service, "North Rim."

Fewer than three in a thousand: National Park Service, "Stats Report Viewer."

"Explosion" a technical term: Elton, *Ecology of Invasions,* 18, 31.

Far surpasses natural rates of change: U.S. Congress, Office of Technology Assessment, *Harmful Non-indigenous Species,* 78.

One count of free-living aliens: Pimentel et al., *Environmental and Economic Costs.*

Biological roulette: U.S. Congress, Office of Technology Assessment, *Harmful Non-indigenous Species,* 1.

Research confirms that many invasives: Hellmann et al., "Five Potential Consequences."

A factor in nearly half of the cases: Pimentel et al., *Environmental and Economic Costs.*

As invasives overtake the national parks: U.S. Congress, Office of Technology Assessment, *Harmful Non-indigenous Species,* 79.

Invasives and noxious weeds are already the dominant vegetation: Bureau of Land Management, *Federal Rangeland Policy, Draft Programmatic Environmental Impact Statement,* 2005, cited in Donahue, "Federal Rangeland Policy," 301.

Nearly two hundred species of nonnative plants: Lori Makarick, interview.

They eat a lot: Martha Hahn, interview.

About eighty of Grand Canyon's alien plant species: Makarick, interview.

Millions more have altered the banks: IMS Health, "Tamarix Genus Distribution."

The heaviest and most damaging infestations: CAB International, "Invasive Species Compendium."

Tamarisk is efficient at pulling water out: Goodwin and Burch, "Watch out for Saltcedar."

Show up in the food chain: Eisler, *Mirex Hazards to Fish,* 2, 13–18.

Outrageous economic and ecological costs: Niemelä and Mattson, "North American Forests," 751.

Orders of magnitude greater: Pimentel et al., *Environmental and Economic Costs,* 1–16.

Annual cost of invasives: Pimentel et al., "Update," 273.

A federal survey of several agencies: U.S. General Accounting Office, *Invasive Species,* 7.

$2.3 trillion in goods: U.S. Department of Commerce, "U.S. International Trade."

Strategies have long been proposed: P. Jenkins, "Paying for Protection."

The federal government itself defines regulatory capture: U.S. Government Accountability Office, *Federal User Fees,* 22.

More than 2,200 different kinds of alien wild animals: P. Jenkins, "Failed Regulatory System," in Witmer, Pitt, and Fagerstone, "Managing Vertebrate Invasive Species," 86.

Twenty-foot-long giant rock python: Dell'Amore, "Python 'Nightmare.'"

The tumbleweedy Russian thistle: Joshua Higgins, interview.

Chapter Three. Landscapes in Motion

An enormous leap: Douglas Schwartz, interview.

Skeletal remains from the region: No evidence of violent conflict has materialized around the Canyon itself, Douglas Schwartz told me, but that is not surprising, considering the sparse physical remains. In the larger region, however, the evidence of violent conflict in times of climate pressure and famine is widespread. See Leblanc, *Prehistoric Warfare.*

Food storage structures: Other archaeologists have theorized instead that these were watchtowers or defensive structures, according to Jan Balsom, interview.

A hundred thousand years: Archer, *Long Thaw,* 11; Bernstein et al., *Climate Change 2007,* 47.

Moderate to extreme drought continues: "SW Climate Outlook."

By 2017 it had dropped to its lowest elevation: R. Davis, "2015 Lake Powell Water."

Widespread tree death and fires: Garfin et al., "Southwest," in Melillo, Richmond, and Yohe, *Climate Change Impacts,* 463–64.

Temperatures in the Southwest are projected to rise: Ibid., 454.

If, instead, greenhouse gas emissions: Ibid.

Ecological stress in the region's forests: Park Williams, interview.

Well over a thousand years: Ibid.

Rare Mexican spotted owl: U.S. Fish and Wildlife Service, *Recovery Plan,* 298.

Gambel's oak, juniper, and piñon pine: Rehfeldt et al., "Empirical Analyses," 1131–33.

At Sequoia National Park: Nathan Stephenson, interview.

Glaciers at Glacier National Park: Nash, "Twilight of the Glaciers" Adema, Karpilo, and Molnia, "Melting Denali."

At Great Smoky Mountains: Burns, Johnston, and Schmitz, "Global Climate Change."

Fifty-four plant and animal species at Hawai'i Volcanoes National Park: Rhonda Loh, interview.

Trend is nearly as grim: Burns, Johnston, and Schmitz, "Global Climate Change."

84 percent of park units: Gonzalez, "State of Science."

Forest ecologist Hank Shugart: Nash, *Virginia Climate Fever,* 64.

There's no agreement on it: Ibid., 64–65.

Achilles' heel for species preservation: Burns, Johnston, and Schmitz, "Global Climate Change," 11474.

Chapter Four. Ghost Tour

Caves in the Redwall limestone: Janice Stroud-Settles, correspondence.

Condor hatchlings are all too rare: Chris Parish, interview.

Might just as well be called the Grand Canyon wolf: We don't yet know, and may never know, which subspecies of wolf inhabited the Kaibab Plateau and what is now Grand Canyon National Park. The question is of interest to scientists and matters in the interpretation and implementation of the Endangered Species Act. It matters little to wolf restoration at the park. Southern subspecies—"Mexican"—and northern gray wolves were part of a large, continuous population adapted to survive well in its more northerly or southerly circumstances, and they were very likely to have hybridized freely, blending along a continuum of what are now subspecies. The population ecologist and geneticist

Richard Fredrickson writes,

> To my knowledge it is not clear what type of wolf inhabited the Grand Canyon
> area in recent historical times. No historical specimens from the area have been
> genotyped. In the broad region of the southern Rockies there is evidence that this
> larger area was a mixing zone between *C.l. nubilus* (northern gray wolves) and
> "southern clade" wolves (which include and may be synonymous with Mexican
> wolves). In my opinion, it is possible that some Mexican wolves may have at times
> occupied the GC area, but my guess is that most (or perhaps all) wolves in the
> area were C.l. nubilus, nubilus × Mexican Wolf hybrids, or *nubilus* with some
> small amount of Mexican Wolf ancestry. The northern limit of Mexican Wolf
> historic range has been considered to be substantially south of the Grand Canyon.
> But relying solely on naturalists' observations to determine "historic range" is not
> necessarily a very reliable or precise way of defining it. (correspondence)

Trapped, poisoned, or shot: Ripple and Beschta, "Linking Wolves and Plants."
Grizzly bears may have been here: D. Brown, *Grizzly in the Southwest,* 79; David
Mattson, interview.
Last of the jaguars: Lange, "Jaguar in Arizona."
Habitat of playful, mercurial southwestern river otters: Paul Polechla, interview.
Pushed to that edge: William Newmark, interview.
Disappeared after those parks were established: Newmark, "Land-Bridge Island
Perspective."
If the political side was cleaned up: John Vucetich, interview.
Largely paid for by private groups: Parish, interview.
22 condors remained on the planet: Walters et al., *Status of the California Condor,* 2.
6 birds were released in 1996: National Park Service, "Condor Re-
introduction."
70 condors soar over Arizona: National Park Service, "California Condors in
AZ/UT."
Focused labor . . . tens of millions of dollars: John McCamman, correspondence.
Condor 133 was hatched . . . last living member: Chris Parish, correspondence.
Couple of million years of endurance: Chamberlain, Waldbauer, and Fox-Dobbs,
"Pleistocene," 16707.
They fall prey to coyotes: Rideout et al., "Patterns of Mortality," 95; Parish,
interview.
Their worst enemy . . . killed by lead poisoning: Parish, interview.
Snowstorm effect: Jeffrey Walters, interview.
If they ingest lead: Center for Biological Diversity v. United States Forest
Service, No. 13–16684, DC, No. 3:12-CV-08176-SMM (9th Cir. Nov. 18, 2015).
Digestive system . . . horrible way to die: Janice Stroud-Settles, interview.

Sometimes Arizona even argues: Martha Hahn, correspondence.

The Forest Service does have that authority: Center for Biological Diversity v. Forest Service, 3:12-CV-08176-SMM.

Their recovery as a wild population is completely stymied: Walters, Derrickson, and Fry, *Status of the California Condor.*

More birds mean more staff: Walters, interview.

Some programs in the United States: Ibid.

Chiseled out of the northern walls: Dollar, *Guide to Arizona's Wilderness,* 54–55.

A federal campaign on this plateau: Rasmussen, "Biotic Communities," 236.

Now there is abundant evidence: Ripple and Beschta. "Wolves and the Ecology."

Nearly ten thousand cattle graze the Kaibab: Michael Hannemann, correspondence.

Herd now estimated at 88 million: Plain and Brown, "Annual Cattle Inventory Summary."

One female gray wolf. . .shot by a hunter: U.S. Fish and Wildlife Service, "Investigation Complete."

Ranged through two-thirds: U.S. Fish and Wildlife Service, *Draft Mexican Wolf,* 33.

Among the first waves of wolf migrants: Robert Wayne, correspondence.

Began a captive breeding program: Fredrickson, interview.

A court ruling finally forced the release: U.S. Fish and Wildlife Service, "Mexican Wolf Recovery Timeline."

After some forty years: Bradley and Lambert, "2015 Mexican Wolf Population."

$30 million by the most recent estimate: U.S. Fish and Wildlife Service, *Mexican Wolf Blue Range.*

The Fish and Wildlife Service is the lead agency: Kirsten Leong, interview.

Our neighbors don't want wolves: David Uberuaga, interview.

A 2005 survey: Social Research Laboratory, *Grand Canyon Wolf Recovery,* 2–3.

A more recent survey: Research and Polling, "Wolf Recovery Survey," 1–22.

The latest in a long series of lawsuits: U.S. Fish and Wildlife Service, "Mexican Wolf Recovery Efforts."

The governors of Colorado, Utah, Arizona, and New Mexico: "Where's the Money?"

An instructive list of objections: Washington County, Utah, Commission, Resolution R-2012–1610, "A Resolution to Adopt a Policy Opposing the Introduction of the Mexican Gray Wolf," January 17, 2012.

Wolves do prey: U.S. Fish and Wildlife Service, "Environmental Impact Statement," 4:24.

Reintroduction of wolves at Yellowstone: William Ripple, interview.

The main stem of Zion Canyon: Ripple and Beschta, "Linking a Cougar Decline."

Fishing and hunting in Utah declined: U.S. Fish and Wildlife Service, U.S. Department of Commerce, and U.S. Census Bureau, *National Survey of Fishing.* Cf. U.S. Fish and Wildlife Service, U.S. Department of Commerce, and U.S. Census Bureau, *1996 National Survey.*

Not just the ones with guns: "Help Mexican Wolves Recover."

There is no evidence that wolves pose much of a threat: Fredrickson, interview.

Later, a local congressman submitted a bill: Goad and Santschi, "Baca Bill Aims."

Tapestries of natural systems and other species: All details on Delhi Sands are from Greg Ballmer, correspondence.

Eighty-one legislative attacks: Center for Biological Diversity, "Attacks."

Interest groups such as the American Petroleum Institute: Snow, "House Approves Bill."

Ryan Zinke of Montana: Center for Biological Diversity, "Rep. Zinke's Vote Record."

The wild population fell back: U.S. Fish and Wildlife Service, "Mexican Wolf Recovery Area."

Illegal shooting or trapping: U.S. Fish and Wildlife Service, *Mexican Wolf Recovery Program,* 33.

Managers should actively solicit input: Goldman et al., *Progress and Problems,* 5, 7.

The Service is fully committed to the highest standards. Vanessa Kauffman, correspondence.

The jaguar's historical range: U.S. Fish and Wildlife Service, *Recovery Outline,* 18.

Population of jaguars persists: Ibid.

The agency acknowledged in 1997: Ibid.

John Roll, ruling on the next lawsuit: Center for Biological Diversity v. Dirk Kempthorne, Secretary of the Interior, et al. and Defenders of Wildlife v. Dale Hall, Director, U.S. Fish and Wildlife Service, et al., No. CV-07-372-TUC JMR (Consolidated with Case No. CV-08-335 RCC) (D. Ct. Ariz. March 31, 2009).

At least the habitat was marked out: U.S. Fish and Wildlife Service, "Service Designates Jaguar."

Overruling the findings of his agency's own biologists: T. Davis, "Biologists Who Warned."

Two-thirds of the mine's total of eight square miles were to be on Coronado National Forest land: Steven Spangle, U.S. Fish and Wildlife Service, Arizona Ecological

Service Office, Phoenix, to Jim Upchurch, Forest Supervisor, Coronado National Forest, May 16, 2014.

The initial capture was carefully planned: All details on this episode are from Wagner, "Macho B."

Interest in the future of the jaguar: Povilitis, "Recovering the Jaguar."

Three are publicly avowed members: Kurt Davis, correspondence. The biographical notes of James Zieler and Robert Mansell on the Game and Fish Commission website state that they are NRA members (Arizona Game and Fish Department, "Commission Members"). Jim Ammons is not an NRA member. Edward Madden could not be contacted.

Plain intent of Congress: Center for Biological Diversity v. Dirk Kempthorne, CV-07–372-TUC JMR.

This means that as long as a small population: Vucetich and Nelson. "Conservation, or Curation?" See also the agency's response, in Ashe and Sobeck, "Using Scarce Resources."

Today, the return of the bald eagle: Ibid.

Chapter Five. Tusayans

As long as twenty thousand years: Donald Bills, interview.

About nine hundred seeps and springs: Cynthia Valle, correspondence.

Five hundred times more different species: National Park Service, "Study of Seeps"; Jan Balsom, interview; Ellen Brennan, interview.

Uninhabited for eight centuries: Balsom, interview; Brennan, interview.

In 2003, however, the state legislature amended the law: AZ Rev. Stat. § 9–101.

Five of its seven members said privately: Belson, "Tiny Grand Canyon Town."

Locals received cash "win bonuses": Cole, "Tusayan 'Bonuses' Follow Election."

Affinities of the town council members: Town of Tusayan, "Town Leadership"; Eric Duthie, interview.

Third largest in America: Jesse Tron, correspondence.

Water consumption … would quadruple: Betz, "Park Service to Tusayan"; David Uberuaga, interview.

Third busiest airport in Arizona: U.S. Federal Aviation Administration, *Calendar Year 2014.*

Backlog of $371.6 million: National Park Service, "Deferred Maintenance" (2015); Bruce Sheaffer, interview.

Another commercial thrust: Confluence Partners, "Our Land."

"The dirty little secret": Kling, "Preserving Yellowstone."

Grand Teton National Park took decades: Skaggs, "Creation of Grand Teton."

Freeman Tilden . . . addressed the error: Freeman Tilden, qtd. in Loomis, "Congressional Testimony."

Walt Disney Company put forward plans: Reuters, "Thousands Protest Disney."

Disney halted the project: Perez-Pena, "Disney Drops Plan."

Recent national poll, echoing many others: Hart Research Associates, "Strong Bipartisan Support."

Chapter Six. Destitution Park

Martha Hahn, science administrator: Martha Hahn has since retired.

Burdened with fighting accelerating waves: U.S. Forest Service, *Rising Cost.*

10 percent fewer employees: Linda Smith, correspondence.

Oil and gas extraction royalties: U.S. Bureau of Labor Statistics, "CPI Inflation Calculator" (figures exclude fire management); U.S. Office of Natural Resources Revenue, "Reported Revenues."

Overextended and underfunded: Maza, "Secretary Jewell."

Had not been inspected for safety: Yen and Peipert, "Oil and Gas Wells."

They are already replacing it: Shedlowski, "Grand Canyon."

About six hundred fewer full-time employees: National Park Service, "Fiscal Year 2015."

Totals $371.6 million: National Park Service, "Deferred Maintenance" (2015); Bruce Sheaffer, interview.

British Petroleum oil spill: Nash, "Oil and Water."

$12 billion to $47 billion in annual benefits: Ibid.

Conservation biologist Michael Soulé: Ibid.

Whether talking to conservatives or liberals: Metz and Weigel, *Language of Conservation.*

One would be hard pressed: Tennessee Valley Authority v. Hill, 437 U.S. 153 (1978).

Committee some referred to as the "God squad": U.S. Congress, "Endangered Species Act Amendments."

Park's overall budget: Thomas Shehan, interview and correspondence.

Prairie dogs, swift foxes, burrowing owls: Glenn Plumb, interview.

Trajectory of potential restoration: Gary Machlis, interview.

Interns just to establish a baseline for insects: David Uberuaga, interview.

Given enough time, some of these reports: Chester, "Revisiting Leopold."

Environmental changes . . . widespread: National Park Service, *Revisiting Leopold,* 4–5.

38 percent of visitors to Grand Canyon: Uberuaga, interview.

$15.7 billion into the cash registers: Thomas, Huber, and Koontz, *Spending Effects.*

Twenty-five dollars per flyover: Robin Martin, correspondence.

More than one hundred thousand overflights: National Park Service, "Special Flight Rules Area," 128–29, 135.

AARP does not lobby Congress: Kristin Palmer, correspondence.

2016 budget request: Beverly Stephens, correspondence.

Visitation . . . in 2014 jumped 5 percent: National Park Service, "Annual Visitation Summary Reports."

Oversimple plan: The National Park Service does not currently charge fees at many of the most-visited park units. Great Smoky Mountains National Park, for example, was created in the 1930s by agreement with the states that there would be no fees (they could be invited to reconsider, of course). At other park units collecting fees is difficult because there are multiple points of entry. The hugely popular Blue Ridge Parkway is heavily used by commute traffic as it passes near larger cities. Also, this plan would mean changing the rules—some imposed by the park system, some from elsewhere—about how fee money can be spent. But the parks that now charge an entry fee attracted more than 123 million visitors in 2015 and collected a pittance. Whatever its caveats, my suggestion at least invites fresh thinking for cash-strapped public lands. Half of 123 million—our fudge factor—is 61.5 million. Multiply that by twenty dollars per person and the parks collect $1.23 billion. Deduct the $200 million—probably less—that was collected in 2015 under the current system. That leaves the billion-dollar bonus cited in the text. The figures on visitors and receipts are from Jessica Bowron and Beverly Stephens of the National Park Service budget office.

Chapter Seven. Air and Uncle John

Whole damn story: Gerke, "National Park Service Interpretation."

Park Service air-quality specialist: Shannon Reed has since left this position.

Thirteen thousand or so people: National Park Service, "Recreation Visitors by Month."

A quarter of the distance: Shannon Reed, interview.

Natural visual range: National Park Service. "Air Pollution Impacts."

Impaired 90 percent of the time: Environmental Protection Agency, "New Pollution Controls"

Words of a National Research Council study: National Research Council, *Protecting Visibility,* 1.

John Muir extolled: Muir, "Grand Cañon of the Colorado," 108.

Fair and reasonable demand: Georgia v. Tennessee Copper, 206 U.S. 230 (1907).

Congress created the National Park Service: Dilsaver, *America's National Park System,* ch. 1.

Anguished resignation letter lamented: Fiore, "Top EPA Enforcement Official."

No matter what long-term promises are made: The Obama administration launched a national Clean Power Plan in 2015, for example, calling for broad cutbacks, to be implemented over fifteen years, on carbon dioxide emissions. As of 2017, more than half the states are suing the EPA over the plan, and its fate is uncertain.

Visibility has improved marginally by some measures: Bret Schichtel, interview.

Largest in the West: Synapse Energy Economics, "Renewable Energy Alternatives," 1.

Still accounts for most of the haze: Environmental Protection Agency, "Approval of Air Quality," p. 9, table 43, p. 130; John Vimont, interview.

Warned by federal scientists in advance: Lustgarten, "Miracle Machines"; National Public Radio, "Historical Blunder."

78-mile-long dedicated electric rail line: Scott Harelson, interview.

Fifteen million tons of carbon dioxide: Environmental Protection Agency, "Emissions."

Nine billion gallons: Hurlbut et al., "Navajo Generating Station," 24.

2,900 feet up: Central Arizona Project, "CAP Background"; Arizona Department of Water Resources, "Active Management Area."

Surplus in the region: Hurlbut et al., "Navajo Generating Station," iii, iv.

A lot of people would be out of work: Hardeen, "Black Mesa Railroad."

Can reduce the immune system's ability: Environmental Protection Agency, "Written Report," 12.

Hourly readings at the Abyss: John Vimont, correspondence.

Human toll from nitrogen-oxide: Environmental Protection Agency, "Written Report."

Exacerbate the harms: Ibid., 12.

Navajo ranks third: Enesta Jones, correspondence (ranking is for 2014).

Several hundred pounds of mercury: Environmental Protection Agency, "Emissions."

Warning about high levels of mercury: Arizona Game and Fish Department, "Fish Consumption Advisories."

Mercury concentrations in insects: Walters et al., "Mercury and Selenium Accumulaåtion," 1–10.

Fantastic solar resources: Paul Denholm, interview.

Hundred square miles with solar panels: Denholm, interview; Ong et al., *Land-Use Requirements,* 19, 24, 36; Christopher Namovicz, interview.

Initial costs and a payback schedule: According to U.S. Bureau of Labor Statistics. "CPI Inflation Calculator," the $670 million original cost in 1974 was $3.2 billion in 2015 dollars.

Coal mining is a short-term crutch: Johnson, "Navajo Nation."

Wind and solar farms: Synapse Energy Economics, "Renewable Energy Alternatives."

Overall average is 24 percent: Morris, "Renewables Briefly Covered"; Conca, "Germany's Energy Transition"; Salt River Project, "Facts about SRP."

Multipurpose federal reclamation project: Paul Ostapuk, interview.

Mining the sun, and harvesting the wind: Synapse Energy Economics, "Renewable Energy Alternatives," 10.

Chapter Eight. Mount Trumbull

Described as one of the most remote: Goolrick, *Parashant National Monument,* iii.

Administers some 383,000 square miles: Bureau of Land Management, "Bureau of Land Management."

60 percent leased for grazing: Bureau of Land Management, "Fact Sheet."

Forest Service leases grazing rights: Knowlton, "Home on the Range."

There go the states: "Fast Facts Study Guide."

Thirteen parks, including 1,200 square miles: Ratner, "Service Abandons Its Mission," 12.

More than five hundred years old: Moore, Davis, and Duck, "Mt. Trumbull Ponderosa Pine," 118.

Sawyers had to let them stand: Covington et al., "Changes in Ponderosa Pine," 20.

Build a handsome Mormon temple: Moore, Davis, and Duck, "Mt. Trumbull Ponderosa Pine," 20.

No one really knows: Whit Bunting, interview.

The Ungrazed Piece of Land: Kathleen Harcksen, interview.

Research by David Tilman: Tilman, Reich, and Knops, "Biodiversity and Ecosystem Stability."

Average is only five inches: Jepson el al., "Temperature and Precipitation Data," 40.

Daddy said that when they first moved: National Park Service, "Journey through Time."

Evidence that more diverse vegetation does return: McGlone, Springer, and Laughlin, "Pine Forest Restoration"; McGlone et al., "Nonnative Species Influence"; McGlone, Sieg, and Kolb, "Invasion Resistance and Persistence."

Monument is a model: National Park Service, "Virtual Visitor Center."

As many as 3,500 animals: Rachel Carnahan, correspondence.

Pulverize its nitrogen-fixing biological crusts: Jayne Belnap, interview; Matthew Bowker, interview.

Few buffalo west of the Rockies: Thomas Fleischner, interview.

Wealth of research affirms: See, for example, Reisner et al., "Conditions Favouring *Bromus tectorum.*"

Cattle's hooves conveniently press seeds: Harcksen, interview.

After years of toilsome research: McGlone et al., "Nonnative Species Influence." For the role of the cattle, see 200–201.

Cheatgrass can erase native perennials: McGlone, Springer, Laughlin, "Pine Forest Restoration," 2638, 2644.

We like cheatgrass: Bill Bundy, interview.

If cattle grazing is … promoting cheatgrass: Sorensen, "Ponderosa Pine Understory Response," 125.

It's been a learning experience for us: David Huffman, interview.

Condition of the range and vegetation: Bureau of Land Management, Grand Canyon, *Range and Vegetation,* 20.

List of the challenges: David Pyke, interview.

Sampling process can mislead: Richard Mayberry, correspondence.

A couple of dozen research findings: Fleischner, "Ecological Costs."

Grazing may be the major factor: Beschta et al., "Adapting to Climate Change," 477.

Bird populations damaged by grazing: Brennan and Kuvlevsky, "North American Grassland Birds," 1.

Conservation groups are collaborating with ranchers: The conservation groups are The Nature Conservancy and the Grand Canyon Trust.

Tortoises thrive on cow dung: Salazar, "Report."

Tortoise research experts: Brian Wooldridge, correspondence; Roy Averill-Murray, correspondence.

Shifting the burden of proof: Beschta et al., "Adapting to Climate Change," 486.

Legal obligation to discontinue grazing: Debra Donahue, interview; Donahue, "Federal Rangeland Policy," 347.

Chapter Nine. Cash Cows

Challenged by a mining company: Peck, "Controversy," 41–42.

About one percent of the Parashant: Rachel Carnahan, correspondence.

Ranchers and their political allies . . . objected: Findley, "Monumental Betrayal."

Babbitt . . . told a Flagstaff public hearing: McGivney, "Grand Canyon–Parashant."

Angry ranchers who heckled: Ibid.

Babbitt gloated and rocked: Findley, "Monumental Betrayal."

Twenty-eight grazing allotments: Whit Bunting, interview.

I feel like a Phoenix homeowner: Rushlo, "Monument Plan."

Children, too, are often raised to think: Debra Donahue, interview; Kathleen Harcksen, interview.

The most trugit, the most impressive lobbyists: Foss, *Politics and Grass.*

The new federal mission: Bureau of Land Management, "Taylor Grazing Act."

They even offered to leave the room: Foss, Politics and Grass.

Clock is ticking: Bureau of Land Management, "Great Basin."

Agency ought to be induced: Babbitt, "Grand Staircase."

Accused by their own employees: Lipton, "Drillers in Utah."

Actually run fewer cattle: Mark Wimmer, interview.

BLM personnel in charge of monitoring: Bunting, interview.

Ample authority to reduce, revoke, or suspend: Debra Donahue, correspondence.

Prevent unnecessary or undue degradation: Ibid.

1 percent of U.S. beef production: Tom Gorey, correspondence; Plain and Brown, "Annual Cattle Inventory Summary."

Grazing costs the BLM: Gorey, correspondence.

Losses from both Forest Service and BLM administration: Glaser, Romaniello, and Moskowitz, *Costs and Consequences,* 1.

Three large casinos: In Las Vegas, Bellagio employs 9,700; MGM Grand, 9,000; and Mandalay Bay, 8,000. *Casino City Times,* "Top-10 Casinos."

Employers initiated 6,500 layoffs: U.S. Bureau of Labor Statistics, "Report 1043."

Picture available to us from BLM data: Bureau of Land Management, Denver, "Ad Hoc Report." The actual number of square miles for each ownership may be greater than listed here, since BLM records are not always clear regarding multiple ownership arrangements for the same grazing allotment.

Folks leasing great swaths of public land: Forbes, "#382 Bruce R McCaw"; *Forbes,* "#913 Barrick Gold"; *Forbes,* "#58 Stanley Kroenke"; *Forbes,* "#80 Ann Walton

Kroenke"; *Forbes*, "21. W. Barron Hilton"; Markoff, "William Hewlett Dies at 87"; Bureau of Land Management, Denver, "Ad Hoc Report."

Southern Nevada Water Authority holds: Bronson Mack, correspondence.

Forest Service allows grazing: Matt Reeves, interview.

The top 200 permittees: U.S. Forest Service. "Response."

Power's study analyzed: Power, "Taking Stock."

Nearly a dollar cheaper: Gorey, correspondence; U.S. Bureau of Labor Statistics, "CPI Inflation Calculator."

Less than a fifth of what ranchers would pay: Gorey, correspondence.

Chapter Ten. Sacred Cowboys

I can see a BLM range allotment: Donahue, *Western Range Revisited*, ix.

Again took offense on behalf of ranchers: Wyoming State Legislature, "Original Senate Engrossed File."

Headline on an op-ed: Guo, "Wyoming."

David and Charles Koch ... hold ... grazing permits: Bureau of Land Management, Denver, "Ad Hoc Report."

Preserving the cowboy way: Koch Industries, "Matador Cattle Company."

Worth about $35 billion apiece: Alexander, "Turbulent Markets."

Repeatedly asking for government help: Hawkins, "40 Best Political Quotes."

Real power in cultural symbols: Thomas Power, interview.

Real range of subsidies: Donahue, "Western Grazing."

BLM awards the permit: Bureau of Land Management, "Grazing Permit."

Program killed the following: Steve Kendrot, interview; U.S. Department of Agriculture, "Program Data Report G."

More fully understand ecological conditions: Bureau of Land Management, "Rapid Ecoregional Assessments."

Protests from some of the scientists: Bureau of Land Management, *Colorado Plateau Rapid Ecoregional*.

At least two hundred violent attacks: Public Employees, "Employee Violence Report."

Whenever I went to southern Utah: Glionna, "Protesters in Utah."

Helpfully included a form letter: Rampton, "Stop the BLM."

Six million years ago: Taylor Edwards, correspondence.

Tortoise populations ... down 32 percent in just ten years: Allison, *Range-Wide Monitoring*, 6.

Grazing a "major threat": Brian Wooldridge, interview.

Despite requests from the Fish and Wildlife Service: Ibid.

Similar, if less vivid, terms: Leshy and Mcusic, "Where's the Beef."

Harcksen seconds the motion: Kathleen Harcksen, interview.

Allotment data from the BLM: Bureau of Land Management, Denver. "Ad Hoc Report."

Oak Spring, on the southern Parashant: Larry Stevens, interview.

Thirty pounds of vegetation: Rasby, "Beef Cow Consumes"; Rasby, "Cows Drink Per Day."

Wreckage of house trailers: Stevens, interview.

Successes here have been emulated: MacEachern, "America's Best Idea."

Patricia Wilson Clothier's family: Clothier, *Beneath the Window*, 144–46.

BLM Another Intrusive Tyranical: USA Today, "Militia Leader."

Sole juror who spoke to the media: Bernton, "Malheur Case Prosecutors."

At the time he lost his permit: Rob Mrowka, interview.

I don't recognize the United States government: Hayes, "FOX News and GOP."

BLM sent Bundy a Trespass Notice: United States v. Bundy, CV-S-98-531-JBR (RJJ), November 3, 1998.

Notwithstanding the Court's Orders: Ibid.; United States v. Bundy, CV-0804-LDG-GWF, July 9, 2013.

Everyone up to the Secretary of the Interior: Ralston, "Letter from Nevada."

These people are thieves: Sonner and Griffith, "Tension Growing between Ranchers."

Sean Hannity, who spotlighted Bundy: Fox News, "Rancher Cliven Bundy Interview."

About four hundred self-styled patriots: KLAS-8 News, "FBI Investigating Supporters," at 1:25.

Scott Shaw ... explained: Gabbee. "Oklahoma Militia of 50,000."

Tom Jenkins told an interviewer: KLAS-8 News, "FBI Investigating Supporters," at 1:25.

Wanted nothing to do with Bundy's tactics: Rogers, "Nevada Cattlemen's Association."

We were actually strategizing: Fox News, "Richard Mack."

Guided from the top levels: Ralston, "Letter from Nevada."

Account of Ryan Payne: Lenz and Potok, *War in the West*, 14.

All we got to do is open those gates: Ibid.

Armed federal officers suddenly withdrew: Ibid.

Bundy would confide to a political gathering: K. Jenkins, "Bundy."

Ranchers near Battle Mountain, Nevada: Turkewitz, "Drought Pushes Nevada Ranchers."

Two Oregon men with plans: Moriarty, "Sugar Pine Mine Co-owner."

The road into Recapture Canyon: Glionna, "Protesters in Utah."

If things don't change, it's not long: Lenz and Potok, *War in the West,* 15.

Lyman and one other rider: Romboy, "San Juan County Commissioner"; Taylor, "Utah Official Convicted."

Survey of local law-enforcement officials: Kurzman and Schanzer, *Law Enforcement Assessment,* 3.

Malheur occupiers sent out a call: "Oregon Militia Leader."

Country music video: Canady, "Big Park."

Insurrection is not new: This summary of the two insurrections is drawn from Chernow, *Washington.*

A Facebook link, "Shootwelfareranchers": Bennett, "Washington County Commissioner."

Chapter Eleven. Treasure Maps

A sideshow, sometimes applauded: Healy and Johnson, "Quieter Than Bundy." Jennifer Fielder, president of the American Lands Council, posted this approval of the Malheur occupation on January 27, 2016: "Afterwards, about 15 armed protesters occupied vacant office buildings at a nearby wildlife refuge. It was an act of civil disobedience, led by cowboys and backed by well-armed military veterans, that drew worldwide attention. The determined protesters said they wanted the Hammonds released from prison, and control of the federally managed lands turned back over to the people in the area, or to the local government. Why would they risk their lives to make such demands?" (Fielder, "Update"). See also Stein, "Koch Brothers Group."

Arizona legislature, for example, approved this bill: Legiscan, "Bill Text."

More common formula: Price, "Lawsuit over Public Lands."

Legislatures in Utah: Goad and Kenworthy, "State Efforts."

Utah legislature authorized: Romboy, "Herbert Signs Bill."

Republican Party platform: Republican National Committee, "America's Natural Resources."

23,000 active oil, gas, and coal leases: Hein, *Harmonizing Preservation,* 4.

What you pay for a cup of coffee: Ibid.

$300,000 to Zinke's one congressional campaign: Maplight, "U.S. Congress."

During the Obama years: Murphy, "Largest Private Companies."

Ten wealthiest individuals: Alexander, "Turbulent Markets."

Koch Industries employs: Koch Industries, "Life Made Better."

Koch Industries subsidiary controls: Bureau of Land Management, "Land and Mineral Legacy."

Koch Industries' U.S. lobbying expenditures: Center for Responsive Politics, "Oil and Gas Industry."

Each brother pledged $75 million: Vogel, "Kochs Put a Price."

Detractors and supporters alike: Mayer, *Dark Money*, 160.

Koch-supported groups: Graves, "ALEC Exposed."

One model bill . . . states that global warming: Hamburger, Warrick, and Mooney, "Tired of Being Accused."

Disposal and Taxation of Public Lands Act runs to 3,700 words: American Legislative Exchange Council, "Disposal and Taxation."

The BLM hopes to participate fully: Bureau of Land Management, "Treasured Landscapes," 8.

Allow land-use decision makers to act: Ibid., 6.

We now know that these large-scale ecosystems: Ibid., 1.

Only three out of the fourteen: Rob Winthrop, correspondence.

Treasured Landscapes didn't materialize: Tom Gorey, interview.

Governor who does not consider: Fischer, "Ducey 'Skeptical.'"

Livestock do not belong in arid deserts: Devlin, "Babbitt."

Final State of the Union message: Obama, "Remarks."

As one biologist put it: Rob Mrowka, correspondence.

Chapter Twelve. Thrill Rides

River of aircraft: National Park Service, "Special Flight Rules Area," 128–29, 135; Robin Martin, correspondence.

Papillon bills itself: Papillon Airways, "Book a Tour."

New $9 million headquarters: Papillon Airways, "Papillon Grand Canyon Helicopters."

Helicopter motored past: Kurt Fristrup, interview.

Move fifty feet closer: Karen Treviño, interview.

Freshman senator told a committee hearing: U.S. Congress, "Peace Garden," 48.

Ten national parks are trying to manage: Treviño, interview.

Hundred or so other parks: Comay, Vincent, and Alexander, *CRS Report for Congress*, 13.

Several dozen national park units: Ibid., 8.

A thousand snowmobiles a day: Ibid., 1.

Nearly all national forests: Ibid., 4.

A long series of studies: Stokowski and LaPointe, *Environmental and Social Effects,* 3; U.S. Government Accountability Office, "Federal Lands," 40.

Obvious to even casual observers: Stokowski and LaPointe, *Environmental and Social Effects,* 3.

More than 80 percent: Government Accountability Office, "Federal Lands," 37, 40.

$165 million is being spent: Tim Rains, correspondence.

Engineering marvel: National Park Service, "Glacier National Park."

Half-million cars: Vanderbilt and Moler, "Saving a National Treasure," 1; Environmental Protection Agency, *Report,* 5.

Five-year period in the early 1980s: Cummings, "Air Crash Kills," A18.

Points with ringing clarity: U.S. Congress, "Peace Garden."

Did not please the industry: Beamish, "McCain Changes Tune."

Endangered peregrine falcons suffer: National Park Service, "Special Flight Rules Area," 75.

Study of 138 different species of birds: Clinton, "Vocal Traits," 1809.

Grazing patterns of bighorn sheep: National Park Service, "Special Flight Rules Area," 7, 121.

Natural quiet is reported: Ibid., 115, 117.

A third of visitors on short hikes: Ibid., 118, 117, 95.

Can human-caused noise … impair memory? Benfield et al., "Anthropogenic Noise."

Causing a significant adverse effect: National Parks Overflight Act of 1987, Pub. L. 100–91 (1987).

Agency cooperated consistently: Hank Price, correspondence.

Preamble was well outside the norm: National Park Service, *Report to Congress,* iv.

Air-tour industry had not been appeased: Beamish, "McCain Changes Tune."

Not limited just to the air tour industry: Wild Wilderness, "Background Information."

The equivalent of detonating a nuclear bomb: Cart, "Controversy over Plans."

Sentence from the existing policy: Barringer, "Top Official Urged Change"; Bill Wade, interview.

I was profoundly shocked: Cart, "Controversy over Plans."

Option to ban all tourist overflights: Rick Ernenwein, correspondence; National Park Service, *Aircraft Management Plan.*

Aerial viewing platform: Ernenwein, interview (he supports the decision, however.)

High-flyers make most of the noise: National Park Service, "Special Flight Rules Area," 94.

FAA agreed to consider reducing: National Park Service, "Clarifying the Definition."

By 2008 quiet had become this formulation: Ibid.

Maneuvers eliminated more and more classes: Treviño, interview.

Economically viable air tour industry: National Park Service, "Special Flight Rules Area," 3.

Industry's formal response: Beamish, "McCain Changes Tune"; Crowell and Moring, "Detailed Comments," iii.

Induce them to adopt quiet technology: "An Act to Authorize Funds for Federal-Aid Highways, Highway Safety Programs, and Transit Programs, and for Other Purposes." Pub. L. 112–141, 126 Stat. 843 (2012).

Flawed Park Service Air Tour Plan: McCain, "Statement by Senator McCain."

Elderly do not make up a large portion: National Park Service, "Special Flight Rules Area," 114.

Alan Stephen, a spokesman: Kelly, "Congress Thwarts Plan."

Air-tour noise about 80 percent of the time: Bell, Mace, and Benfield, "Aircraft Overflights," 66.

Chapter Thirteen. Capture and Corruption

Smuggled busloads of trail-runners: David Uberuaga, interview.

Martin decided to halt the plague: Barringer, "Parks Chief Sets Conditions"; Barringer, "More on Coke's Role."

Wrested the pertinent email traffic: Steve Martin, interview.

Continue to bang on the McCain door: Beamish, "McCain Changes Tune."

Senator had at least been listening: Silverman, "Gift That Keeps."

Dramatic, near miraculous: Ibid.

Elling Halvorson played a key role: Public Citizen, "Bundler"; Christian Hilland, correspondence; Beamish, "McCain Changes Tune."

Reid is air tour supporter: Neubauer and Cooper, "Name to Know."

Until we create the conditions: Lessig, *Republic, Lost.*

In a recent poll: Ibid.

Generates the best return: Black, *Best Way to Rob,* 87.

Corruption that works through: Thompson, *Ethics in Congress,* 125–26.

A gift can be a bribe: Teachout, *Corruption in America.*

An embarrassed Benjamin Franklin: Ibid.

Close kin to bribes: Ibid.

Would that be proper: Lessig, *Republic, Lost.*

Barry Goldwater clarified: 145 Cong. Rec. 25517 (1999). (Tom Daschle, citing Sen. Barry Goldwater).

Questions of honor are raised: McCain, *Worth the Fighting For,* 13.

Freebies like ball point pens: Lieb and Scheurich, "Contact between Doctors."

Likens this to the "rotten boroughs": Teachout, *Corruption in America.*

30 to 70 percent of it: Thompson, *Ethics in Congress,* 2.

Relied on faulty assumptions: McCain, "Statement by Senator McCain."

Form of political corruption: "Regulatory Capture."

FAA spokesperson contradicted: Hank Price, correspondence; Alison Duquette, correspondence.

Addendum seemed to confirm: U.S. Congress, "Report."

This explanation crops up in a "conference report" by those who helped enact the change. Alexis Fetzer explains, "A conference report is considered part of a law's legislative history, not the law itself. Legislative history material has only persuasive legal authority and that level of persuasiveness varies depending upon the court interpreting the legislation and the type of legislative history material. Normally reports by congressional committees that considered the proposed legislation before being passed are considered the best source for determining legislative intent. However, these reports are still only persuasive. If a law can be interpreted in multiple ways, a court is not bound to interpret the law as evidenced by Congress' intent through legislative history. That being said, it is unlikely (but not unheard of) that a court would choose to disregard legislative intent unless some other authority takes precedence (i.e. a higher court's interpretation of the law)."

Inspectors said they had been silenced: Critical Lapses in Federal Aviation Administration Safety Oversight of Airlines: Abuses of Regulatory "Partnership Programs": Hearing before the Committee on Transportation and Infrastructure, House of Representatives, 110th Cong., 2d sess. (2008). All related testimony is from this source. For one of several examples, see the testimony of Charalambe Boutris, aviation safety inspector and Boeing 737–700 partial program manager for Aircraft Maintenance, Southwest Airlines Certificate Management Office.

Planes were crashing far more frequently: Rentz and Young, "Frontline."

So does regulatory capture: Carpenter and Moss, *Preventing Regulatory Capture.*

Now under 20 percent: Pew Research Center, "Public Trust in Government."

Americans should be alarmed: Painter, "Conservative Case."

Whether capture can be avoided: Carpenter and Moss, *Preventing Regulatory Capture.*

Financially enfeebled news media: Sullivan, "Investigative Reporting's Future."

Chapter Fourteen. *The Lost Orphan's Legacy*

Ore was among the purest: National Park Service, "Tonto Trail."

Withdrawal, by the secretary of the interior: U.S. Department of the Interior, "Secretary Salazar Announces Decision."

Based on information received: U.S. Department of Agriculture, "Kaibab National Forest."

After the Forest Service's decision: David Uberuaga, interview.

Colorado was around 80 degrees: National Park Service, "Grand Canyon National Park."

Helped drive out three of the eight native fish: Brian Healy, interview.

Blocks the natural flow of sand: Ibid.

More than a thousand U.S. rivers: American Rivers, "Frequently Asked Questions."

Hundreds of billions of gallons: Powell, *Dead Pool,* 118.

Below the power generating turbine intakes: Snider, "Colorado River."

Downstream at Lake Mead: Barnett and Pierce, "Lake Mead."

Alternative to reasoned solutions: Ibid.

Glen Canyon Institute: Power, Power, and Brown, *Impact of the Loss,* 3.

Dan Beard, commissioner: Beard, *Deadbeat Dams.*

Blue-ribbon Park Service advisory committee: National Park Service, *Revisiting Leopold,* 9.

SOURCES AND BIBLIOGRAPHY

INTERVIEWS AND CORRESPONDENCE

Abbey, Robert. Former director, Bureau of Land Management.

Anderson, Mark. Director of conservation science, The Nature Conservancy.

Averill-Murray, Roy. Desert tortoise recovery coordinator, Fish and Wildlife Service.

Ballmer, Greg. Entomologist, staff research associate, University of California Riverside Cooperative Extension.

Balsom, Jan. Deputy chief of science and resource management, Grand Canyon National Park.

Barnett, Tim. Marine research physicist emeritus, Scripps Institution of Oceanography, University of California–San Diego.

Belnap, Jayne. Research ecologist, U.S. Geological Survey.

Beschta, Robert. Geomorphologist, Oregon State University.

Bills, Donald. Hydrologist, U.S. Geological Survey.

Bowker, Matthew. Assistant professor of forest soils and ecosystem ecology, School of Forestry, Northern Arizona University.

Bowron, Jessica. Chief, Division of Budget Formulation and Strategic Planning, National Park Service.

Brennan, Ellen. Archaeologist, cultural resource program manager, Grand Canyon National Park.

Breshears, David. Lead scientist, Terrestrial Ecology Lab, University of Arizona.

Bundy, Bill. Rancher, Lazy S-O Ranch, northern Arizona.

Bundy, Orvel. Rancher, Lazy S-O Ranch, northern Arizona.

Bunting, Whit. Rangeland-management specialist, Bureau of Land Management.

Carnahan, Rachel. Public affairs officer, Arizona Strip District, Bureau of Land Management.

Cavey, Joseph. Former branch chief, Agricultural Plant Health Inspection Service (retired).

Clark, Roger. Program director, Grand Canyon Trust.

Crumbo, Kim. Director of conservation, Grand Canyon Wildlands Council.

Dean, Jeffrey. Dendrochronologist, professor emeritus, University of Arizona.

Denholm, Paul. Principal energy analyst, National Renewable Energy Laboratory.

DePaolo, Tom. Managing director for U.S. investments, Gruppo Stilo USA.

Dobson, Carey. Rancher, Timberline Ranches, Show Low, Arizona.

Dombeck, Michael. Former chief, Forest Service; former director, Bureau of Land Management (retired).

Donahue, Debra. Professor, College of Law, University of Wyoming.

Duquette, Alison. Spokesperson, Office of Communications, Federal Aviation Administration.

Duthie, Eric. Town manager, Tusayan, Arizona.

Edelman, Norman. Chief medical officer, American Lung Association.

Edwards, Taylor. Assistant staff scientist, University of Arizona Genetics Core.

Ernenwein, Rick. Former senior planner, Grand Canyon National Park (retired).

Falzarano, Sarah. Former geographic information system specialist, Soundscape and Overflight Program, Grand Canyon National Park.

Fetzer, Alexis. Librarian, Reference and Research Services, University of Richmond School of Law.

Fleischner, Thomas. Biologist, director, Natural History Institute; professor of environmental studies, Prescott College.

Fredrickson, Richard. Population ecologist and geneticist consultant, Fish and Wildlife Service.

Fristrup, Kurt. Senior scientist, Natural Sounds and Night Skies Division, National Park Service.

Gorey, Tom. Senior public affairs specialist, Bureau of Land Management.

Hahn, Martha. Former chief of science administration and resource management, Grand Canyon National Park (retired).

Hannemann, Michael. Program manager, Range, Watershed, Invasive Species, and Rare Plants, Kaibab National Forest.

Harcksen, Kathleen. Former assistant manager, Grand Canyon–Parashant National Monument, Bureau of Land Management (retired).

Harelson, Scott. Manager, Media Relations, Salt River Project.

Hayhoe, Katharine. Director, Climate Science Center, Texas Tech University.

Healy, Brian. Program manager, Fisheries, Grand Canyon National Park.

Higgins, Joshua. Biological science technician, crew leader, Invasive Plants Department, Grand Canyon National Park.

Hilland, Christian. Deputy press officer, Federal Election Commission.

Huffman, David. Director of research and development, Ecological Restoration Institute, Northern Arizona University.

Jacobs, Andy. Team member, Policy Development Group, Scottsdale, Arizona.

Jones, Enesta. Spokesperson, Environmental Protection Agency, Office of Media Relations.

Kauffman, Vanessa. Spokesperson, Division of Public Affairs, Fish and Wildlife Service.

Keating, Molly. Program analyst, Budget Formulation Division, Bureau of Land Management.

Kendrot, Steve. Deputy director, Wildlife Services, Agricultural Plant Health Inspection Service.

Kodish, Stephanie. Attorney, National Parks Conservation Association.

Lattin, Peter. Former contractor, Geographic Information Systems; project leader, Rapid Ecoregional Assessments, Bureau of Land Management.

Leong, Kirsten. Social research specialist, National Park Service.

Levy, Laura. Former analyst, Soundscape and Overflight Program, Grand Canyon National Park.

Littlesunday, Jarbison. Control room operator, Navajo Generating Station.

Loh, Rhonda. Natural resource manager, Hawai'i Volcanoes National Park.

Machlis, Gary. Science adviser to the director, National Park Service.

Mack, Bronson. Public outreach spokesperson, Southern Nevada Water Authority.

Makarick, Lori. Vegetation program manager, Grand Canyon National Park.

Martin, Robin. Chief, Office of Planning and Compliance. Grand Canyon National Park.

Martin, Steve. Former superintendent, Grand Canyon National Park (retired).

Masayesva, Vernon. Former member, Hopi Tribal Council.

Mattson, David. Wildlife ecologist, School of Forestry and Environmental Studies, Yale University.

Mayberry, Richard. Rangeland-management specialist, Division of Forest, Rangeland, Riparian and Plant Conservation, Bureau of Land Management.

McCamman, John. California condor coordinator, Pacific Southwest Region, Fish and Wildlife Service.

McCauley, Douglas. Marine biologist, University of California–Santa Barbara.

McMullen, Ken. Former program manager, Soundscape and Overflight Program, Grand Canyon National Park (retired).

Menke, Kurt. Certified Geographic Information Systems professional, Bird's Eye View GIS, Albuquerque, New Mexico.

Moffett, Judy. Engineer, Navajo Generating Station.

Mrowka, Rob. Former biologist, Center for Biological Diversity.

Namovicz, Christopher. Senior renewable-energy analyst, Office of Electricity, Coal, Nuclear, and Renewable Analysis, Energy Information Administration.

Nations, James. Former director, Center for Park Research, National Parks Conservation Association.

Newmark, William. Conservation biologist, research curator, Utah Museum of Natural History.

Noss, Reed. Conservation biologist, University of Central Florida.

Ostapuk, Paul. Environmental and operations manager, Navajo Generating Station.

Palmer, Kristin. Director, Media Relations, American Association of Retired Persons.

Parish, Chris. Supervisor, Condor Field Project, Peregrine Fund.

Plumb, Glenn. Chief wildlife biologist, National Park Service.

Polechla, Paul. Biologist, associate research professor, University of New Mexico.

Povilitis, Tony. Wildlife biologist, Life Net Nature.

Power, Thomas. Economist, former chair of the Economics Department, University of Montana.

Price, Hank. Spokesperson, Office of Communications, Federal Aviation Administration.

Propst, Luther. Lawyer, founder, Sonoran Institute.

Pyke, David. Supervisory research ecologist, Forest and Rangeland Ecosystem Science Center, U.S. Geological Survey.

Rains, Tim. Spokesperson, Media relations, Glacier National Park.

Reed, Shannon. Air quality specialist, Grand Canyon National Park.

Reeves, Matt. Research ecologist, Forest Service.

Ripple, William. Landscape ecologist, Oregon State University.

Rodgers, Jane. Deputy chief of science and resource management, Grand Canyon National Park.

Schichtel, Bret. Research physical scientist, Cooperative Institute for Research in the Atmosphere, National Park Service.

Schwartz, Douglas. Senior scholar, School for Advanced Research.

Sheaffer, Bruce. Comptroller, National Park Service.

Shehan, Thomas. Chief of administration, Grand Canyon National Park.

Shepherd, Don. Environmental engineer, Air Resources Division, National Park Service.

Shugart, Hank. Forest ecologist, University of Virginia.

Sims, William. Town attorney, Tusayan, Arizona.

Smith, Linda. Acting deputy assistant director, Business, Fiscal, and Information Resources, Bureau of Land Management.

Soulé, Michael. Professor emeritus of environmental studies, University of California–Santa Cruz.

Sprouse, Gary. Rancher, real estate developer, oil and gas developer, Ely, Nevada.

Stephens, Beverly. Special assistant to the comptroller, National Park Service.

Stephenson, Nathan. Forest ecologist, Western Ecological Research Center, U.S. Geological Survey.

Stevens, Larry. Evolutionary ecologist, curator of ecology and conservation, Museum of Northern Arizona.

Stroud-Settles, Janice. Wildlife biologist, Grand Canyon National Park.

Sutherland, Ron. Conservation scientist, Wildlands Network, Durham, North Carolina.

Swain, Sharmistha. Geospatial scientist, Texas Tech University.

Thurston, George. Professor of environmental medicine, New York University School of Medicine.

Treviño, Karen. Chief, Natural Sounds and Night Skies Division, National Park Service.

Tron, Jesse. Director of communications and media relations, International Council of Shopping Centers.

Tuggle, Benjamin. Director, Southwest Region, Fish and Wildlife Service.

Uberuaga, David. Former superintendent, Grand Canyon National Park (retired).

Valle, Cynthia. Hydrologist, Grand Canyon National Park.

Vimont, John. Chief, Research and Monitoring, Air Resources Division, National Park Service.

Vucetich, John. Wildlife ecologist, associate professor, School of Forest Resources and Environmental Science, Michigan Technological University.

Wade, Bill. Former superintendent, Shenandoah National Park (retired); chair, Coalition of National Park Service Retirees.

Walters, Jeffrey. Avian ecologist, Department of Biological Sciences, Virginia Polytechnic Institute and State University.

Wayne, Robert. Professor, Department of Ecology and Evolutionary Biology, University of California–Los Angeles.

Williams, Park. Bioclimatologist, Lamont-Doherty Earth Observatory, Columbia University.

Wimmer, Mark. Acting manager, Grand Canyon–Parashant National Monument, Bureau of Land Management.

Winthrop, Rob. Senior social scientist, Bureau of Land Management.

Wooldridge, Brian. Biologist, Desert Tortoise Recovery Program, Fish and Wildlife Service.

Zganjar, Chris. Climate change–applications developer, The Nature Conservancy.

BOOKS, ARTICLES, AND REPORTS

Adema, Guy, Ronald Karpilo Jr., and Bruce Molnia. "Melting Denali: Effects of Climate Change on Glaciers." National Park Service. nps.gov/articles/aps-v6-i1-c2.htm.

Alexander, Dan. "Turbulent Markets Shake Up Ranking of World's Richest People." *Forbes,* March 1, 2016.

Allison, Linda. *Range-Wide Monitoring of the Mojave Desert Tortoise (Gopherus agassizii): 2013 and 2014 Annual Reports.* U.S. Fish and Wildlife Service, Desert Tortoise Recovery Office. 2015.

American Legislative Exchange Council. "Disposal and Taxation of Public Lands Act." alec.org/model-policy/disposal-and-taxation-of-public-lands-act/.

AmericanRivers."FrequentlyAskedQuestions."americanrivers.org/conservation-resources/river-restoration/removing-dams-faqs/.

Anderson, Michael F., Paul Hirt, and Sarah Bohl Gerke. "Nature, Culture and History at the Grand Canyon." Arizona State University/Grand Canyon Association. grandcanyonhistory.clas.asu.edu/history_science_wildlife.html.

Archer, David. *The Long Thaw: How Humans Are Changing the Next 100,000 Years of Earth's Climate.* Princeton, NJ: Princeton University Press, 2008.

Arizona Department of Water Resources. "Active Management Area Water Supply: Central Arizona Project Water." azwater.gov/AzDWR/Statewide Planning/WaterAtlas/ActiveManagementAreas/PlanningAreaOverview /WaterSupply.htm.

Arizona Game and Fish Department. "Commission Members." 2016. www .azgfd.com/agency/commission/members.

———. "Fish Consumption Advisories: Arizona and Utah Announce Fish Consumption Advisory (Striped Bass) for Mercury in Southern Lake Powell." October 25, 2012. azgfd.com/fishing/fishconsumption.

Ashe, Dan, and Eileen Sobeck. "Using Scarce Resources to Save Endangered Species." *New York Times,* September 4, 2014, A27.

AZ Rev. Stat. § 9–101 (1996 through 1st Reg. Sess. 50th Legis). http://law.justia .com/codes/arizona/2011/title9/section9–101/.

Babbitt, Bruce "From Grand Staircase to Grand Canyon Parashant: Is There a Monumental Future for the BLM?" Transcript of remarks at the University of Denver Law School, February 17. Washington, DC: Department of the Interior, 2000.

Barnett, Tim, and David Pierce. "When Will Lake Mead Go Dry?" *Water Resources Research* 44, no. 3 (2008).

Barringer, Felicity. "More on Coke's Role in a Shelved Bottle Ban." *New York Times,* December 1, 2011.

———. "Parks Chief Sets Conditions for Plastic Bottle Bans." *New York Times,* December 15, 2011.

———. "Top Official Urged Change in How Parks Are Managed." *New York Times,* August 26, 2005.

Beamish, Rita. "McCain Changes Tune on Support for Grand Canyon Air Tours: Tour Operator a Major Campaign Supporter." Center for Public Integrity. December 26, 2011. publicintegrity.org/2011/12/26/7698/mccain-changes-tune-support-grand-canyon-air-tours.

Beard, Dan. *Deadbeat Dams: Why We Should Abolish the U.S. Bureau of Reclamation and Tear Down Glen Canyon Dam.* Boulder, CO: Johnson Books, 2015.

Bell, Paul, Britt Mace, and Jacob Benfield. "Aircraft Overflights at National Parks: Conflict and Its Potential Resolution." *Park Science* 26, no. 3 (2009). nature.nps.gov/ParkScience/index.cfm?ArticleID=349.

Belson, Ken. "Tiny Grand Canyon Town Has Its Say on Big Project. *New York Times,* February 17, 2012.

Benfield, Jacob, Paul Bell, Lucy Troup, and Nick Soderstrom. "Does Anthropogenic Noise in National Parks Impair Memory?" *Environment and Behavior* 42, no. 5 (2010): 693–706.

Bennett, Craig. "Washington County Commissioner Sends Urgent Alert to State Cattlemen's Association." *Iron County Today,* February 10, 2016.

Bernstein, Lenny, Peter Bosch, Osvaldo Canziani, Zhenlin Chen, Renate Christ, Ogunlade Davidson, and William Hare. *Climate Change 2007: Synthesis Report.* Geneva: Intergovernmental Panel on Climate Change, 2008.

Bernton, Hal. "Malheur Case Prosecutors Just Failed, Says Juror No. 4." *Seattle Times,* October 28, 2016.

Beschta, Robert L., Debra L. Donahue, Dominick A. DellaSala, Jonathan J. Rhodes, James R. Karr, Mary H. O'Brien, Thomas L. Fleischner, and Cindy Deacon Williams. "Adapting to Climate Change on Western Public Lands: Addressing the Ecological Effects of Domestic, Wild, and Feral Ungulates." *Environmental Management* 51 (2012): 474–91.

Betz, Eric. "Park Service to Tusayan: Where Is the Water?" *Arizona Daily Sun,* February 28, 2014.

Black, William. *The Best Way to Rob a Bank Is to Own One: How Corporate Executives and Politicians Looted the S&L Industry.* Austin: University of Texas Press, 2014.

Botkin, Ben. "Sen. Ted Cruz Talks to the Editorial Board." *Las Vegas Review-Journal* video. December 14, 2015. m.reviewjournal.com/politics/elections/cruz-favors-returning-federal-land-states-video.

Bradley, John, and Lynda Lambert. "2015 Mexican Wolf Population Survey Reveals More Work to Be Done but Strategy Remains Viable." U.S. Fish and Wildlife Service, Southwest Region. February 18, 2016. fws.gov/news/ShowNews.cfm?ref=2015-mexican-wolf-population-survey-reveals-more-work-to-be-done-but-&_ID=35616.

Brennan, Leonard, and William Kuvlevsky. "North American Grassland Birds: An Unfolding Conservation Crisis?" *Journal of Wildlife Management* 69, no. 1 (2005): 1–13.

Brown, David. E. *The Grizzly in the Southwest.* Norman: University of Oklahoma Press, 1985.

Bureau of Land Management. *2016 Budget Justifications.* doi.gov/sites/doi
.opengov.ibmcloud.com/files/uploads/FY2016_BLM_Greenbook_0.pdf.

————. "The Bureau of Land Management: Who We Are, What We Do."
blm.gov/wo/st/en/info/About_BLM.html.

————. *Colorado Plateau Rapid Ecoregional Assessment.* August 10, 2010. peer.org
/assets/docs/blm/11_30_11_Workshop_Meeting_Notes.pdf.

————. "Fact Sheet on the BLM's Management of Livestock Grazing on
Public Lands." blm.gov/wo/st/en/prog/grazing.html.

————. *Fiscal Year 2011 Rangeland Inventory, Monitoring, and Evaluation Report.* blm
.gov/wo/st/en/prog/more/rangeland_management/rangeland_inventory
.html.

————. "The Great Basin: Healing the Land." April 2000. https://archive
.org/stream/greatbasinhealin29borm/greatbasinhealin29borm_djvu.txt.

————. "How to Get a Grazing Permit." blm.gov/nv/st/en/prog/grazing
/how_to_get_a_grazing.html.

————. "Land and Mineral Legacy Rehost 2000 System: LR2000." Department of the Interior. May 16, 2014. blm.gov/lr2000/index.htm.

————. "Oil and Gas Statistics." blm.gov/wo/st/en/prog/energy/oil_and_
gas/statistics.html.

————. "Rapid Ecoregional Assessments." blm.gov/wo/st/en/prog/more
/Landscape_Approach/reas.htm.

————. "The Taylor Grazing Act: Background." blm.gov/wy/st/en/field_
offices/Casper/range/taylor.1.html.

————. "Treasured Landscapes: Our Vision, Our Values." Internal draft,
discussion paper. 2010. http://img.ksl.com/slc/2292/229224/2292241z.pdf.

————. "Vegetation Treatments Using Herbicides on Bureau of Land Management Lands in 17 Western States: Draft Programmatic Environmental
Impact Statement, at ES-1." In Donahue, "Federal Rangeland Policy."

Bureau of Land Management, Denver. "An Ad Hoc Report from the Rangeland Authorization System." March 14, 2016.

Bureau of Land Management, Grand Canyon–Parashant National Monument. *Condition of the Range and Vegetation: Manager's Annual Report, 2013.*

Burns, Catherine E., Kevin Johnston, and Oswald Schmitz. "Global Climate
Change and Mammalian Species Diversity in U.S. National Parks." *Proceedings of the National Academy of Sciences* 100, no. 20 (2003): 11474–77.

Busse, Ryan. "The House Places No Value on an America Treasure—Its Public Lands." *New York Times,* January 17, 2017.

CAB International. "Invasive Species Compendium." cabi.org/isc/datasheet /52503.

Canady, Terry. "Big Park." American Land Rights Association video. August 29, 2009. youtube.com/watch?v=4xvAB6DAhtM.

Carpenter, Daniel, and David Moss, eds. *Preventing Regulatory Capture: Special Interest Influence and How to Limit It*. New York: Cambridge University Press, 2014.

Cart, Julie. "Controversy over Plans for Changes in U.S. Parks." *Los Angeles Times,* August 26, 2005.

Casino City Times. "Top-10 Casinos by Number of Employees." February 11, 2008.

Center for Biological Diversity. "Attacks on the Endangered Species Act." *Politics of Extinction,* February 2016. biologicaldiversity.org/campaigns /esa_attacks/table.html.

———. "Rep. Zinke's Vote Record against Endangered Species." *Politics of Extinction*. January 30, 2017. biologicaldiversity.org/campaigns/esa_attacks /zinketable.html.

Center for Responsive Politics. "Oil and Gas Industry Profile: Summary, 2015." opensecrets.org/industries/indus.php?ind=E01.

Central Arizona Project. "CAP Background." cap-az.com/about-us /background.

Chamberlain, C.P., J. Waldbauer, and K. Fox-Dobbs. "Pleistocene to Recent Dietary Shifts in California Condors." *Proceedings of the National Academy of Sciences* 102 no. 4 (2005): 16707–11.

Chereb, Sandra. "Interior Chief: Bundy Will Be Held Accountable." *Las Vegas Review-Journal,* June 24, 2015. reviewjournal.com/news/bundy-blm /interior-chief-bundy-will-be-held-accountable#sthash.lMAqszoM.dpuf.

Chernow, Ron. *Washington: A Life*. New York: Penguin Books, 2011.

Chester, Charles. "Revisiting Leopold in the National Parks." *Island Press* (blog). October 2, 2012. islandpress.org/node/499#sthash.lunmferu.dpuf.

Clothier, Patricia Wilson. *Beneath the Window: Early Ranch Life in the Big Bend Country*. Houston: Iron Mountain, 2013.

Cole, Cyndy. "Tusayan 'Bonuses' Follow Election." *Arizona Daily Sun,* March 9, 2011.

Comay, Laura, Carol Vincent, and Kristina Alexander. *CRS Report for Congress: Motorized Recreation on National Park Service Lands*. Congressional Research Service. February 8, 2013.

Conca, James. "Germany's Energy Transition Breaks the Energiewende Paradox." *Forbes,* July 2, 2015. forbes.com/sites/jamesconca/2015/07/02

/germanys-energy-transition-breaks-the-energiewende-paradox/#140be7
bd2968x.

Confluence Partners. "Our Land, Our Community, Our Future: Grand Canyon Escalade; FAQs." grandcanyonescalade.com/faqs/.

Covington, Wallace, Thomas Heinlein, Peter Fulé, Amy Waltz, and Judith Springer. "Changes in Ponderosa Pine Forests of the Mt. Trumbull Wilderness." Bureau of Land Management. November 1999. library.eri.nau .edu/gsdl/collect/erilibra/import/CovingtonEtal.1999.ChangesInPonderosa PineForests.pdf.

Crowell and Moring, LLP. "Detailed Comments of the Grand Canyon Air Tour Interests on NPS's Draft EIS and against NPS's Preferred Alternative to Further Restrict Air Tours over Grand Canyon National Park." June 20, 2011.

Cummings, Judith. "Air Crash Kills 25 at Grand Canyon." *New York Times,* June 19, 1986.

Davenport, Crola. "Obama Designates Two New National Monuments, Protecting 1.65 Million Acres." *New York Times,* December 28, 2016.

Davis, Kurt. "My Turn: New Grand Canyon Monument Is a Bad Idea." *Arizona Republic,* September 28, 2015.

Davis, Rose. "2015 Lake Powell Water Release to Lake Mead Will Increase." *Reclamation: Managing Water in the West.* August 13, 2014. usbr.gov/newsroom /newsrelease/detail.cfm?RecordID=47753.

Davis, Tony. "Biologists Who Warned of Harm to Jaguar Were Overruled." *Arizona Daily Star,* August 16, 2014.

Dell'Amore, Christine. "Python 'Nightmare': New Giant Species Invading Florida." *National Geographic News,* September 14, 2009. news.nationalgeo-graphic.com/news/2009/09/090911-pythons-florida-giant-snakes.html.

Devlin, Sherry. "Babbitt: Time's Ripe for Reform." *Missoulian,* March 14, 2003.

Dilsaver, Lary M., ed. *America's National Park System: The Critical Documents.* National Park Service. 1994. nps.gov/parkhistory/online_books/anps /anps_ii.htm.

Dollar, Tom. *Guide to Arizona's Wilderness Areas.* Englewood, CO: Westcliffe, 1999.

Donahue, Debra. "Federal Rangeland Policy: Perverting Law and Jeopardizing Ecosystem Services." *Journal of Land Use* 22, no. 2 (2007): 229–354.

———. "Western Grazing: The Capture of Grass, Ground, and Government." Pt. 1. *Environmental Law* 35 (2005): 721–806.

———. *The Western Range Revisited: Removing Livestock from Public Lands to Conserve Native Biodiversity.* Norman: University of Oklahoma Press, 2000.

Eisler, Ronald. *Mirex Hazards to Fish, Wildlife, and Invertebrates: A Synoptic Review.* Laurel, MD: Fish and Wildlife Service, 1985.

Elton, Charles. *The Ecology of Invasions by Animals and Plants.* Chicago: University of Chicago, 1958.

Environmental Protection Agency. "Approval of Air Quality Implementation Plans: Navajo Nation; Regional Haze Requirements for Navajo Generating Station—Technical Support Document for Proposed Rule Docket Number: EPA-R09-OAR-2013–0009." January 17, 2013. noticeandcomment.com/Documents-Related-to-Tribal-Consultations-8–20–2012-fn-17016.aspx.

———. "Emissions: Unit Level Data Report, Navajo Generating Station, 2015." January 27, 2016. https://ghgdata.epa.gov.

———. *Light-Duty Automotive Technology, Carbon Dioxide Emissions, and Fuel Economy Trends: 1975 through 2016.* December 2015.

———. "List of 156 Mandatory Class I Federal Areas." *Code of Federal Regulations Reference (40 CFR PART 81).* September 2015. epa.gov/visibility/class1.html.

———. "New Pollution Controls for Power Plants in Arizona." July 2, 2012. www3.epa.gov/region9/mediacenter/azfip/. https://archive.epa.gov/region9/mediacenter/web/html/index-3.html.

———. "Regional Haze Regulations: Final Rule." *Federal Register* 64, no. 126 (1999): 35714.

———. "Written Report of George D. Thurston Regarding the Proposed Navajo Generating Plant EPA Rulemaking." Exhibit 5. December 12, 2013.

"Fast Facts Study Guide (State Areas)." *US50.* theus50.com/area.php.

Fielder, Jennifer. "Update: Why the Oregon Uprising?" American Lands Council. January 27, 2016. americanlandscouncil.org/why_the_oregon_uprising.

Findley, Tim. "Monumental Betrayal: Secretary of Interior Bruce Babbitt Declared Himself Too Impatient for the Democratic Process, and Too Important to Wait on the Arizona Delegation; He Went to the President." *Range,* Summer 2000.

Fiore, Faye. "Top EPA Enforcement Official Quits, Blasts Bush Policy." *Los Angeles Times,* March 1, 2002.

Fischer, Howard. "Ducey 'Skeptical' about Man-Made Climate Change." *Arizona Capitol Times,* April 30, 2015.

Fleischner, Thomas. "Ecological Costs of Livestock Grazing in Western North America." *Conservation Biology* 8, no. 3 (1994): 629–44.

————. "Livestock Grazing and Wildlife Conservation in the American West: Historical, Policy and Conservation Biology Perspectives." In *Wild Rangelands: Conserving Wildlife While Maintaining Livestock in Semi-Arid Ecosystems*, edited by Johan DuToit, Richard Kock, and James Deutsch, 235–65. Hoboken, NJ: Wiley-Blackwell, 2010.

Forbes. "#382 Bruce R McCaw." *Forbes 400*. 2016. forbes.com/lists/2005/54/RUIT .html.

————. "#58 Stanley Kroenke." *Forbes 400*. 2016. forbes.com/profile/stanley-kroenke/.

————. "#80 Ann Walton Kroenke." *Forbes 400*. 2016. forbes.com/profile /ann-walton-kroenke/.

————. "#913 Barrick Gold." *The World's Biggest Public Companies*. May 2016. forbes.com/companies/barrick-gold/.

————. "21. W. Barron Hilton." *The Fifty Biggest Givers*. 2012. forbes.com /pictures/mjh45edhfh/21-william-barron-hilton/?view=pc#2e3103db1a77.

Foss, Phillip. *Politics and Grass: The Administration of Grazing on the Public Domain*. New York: Greenwood, 1960.

Fox News. "Rancher Cliven Bundy Interview with Sean Hannity." YouTube video. Posted by TeaPartyTyme. April 9, 2014. youtube.com/watch?v=YtpUp MdigWk.

————. "Richard Mack on Standoff Strategy: 'Put All the Women Up at the Front.'" YouTube video. Posted by Southern Poverty Law Center. April 14, 2014. youtube.com/watch?v=cllweBGr H3Ak.

Francis, Clinton. "Vocal Traits and Diet Explain Avian Sensitivities to Anthropogenic Noise." *Global Change Biology* 21, no. 5 (2015): 1809–20.

Gabbee. "Oklahoma Militia of 50,000 Stands beside Bundy Ranch against the Feds." YouTube video. April 21, 2014. www.youtube.com/watch?v= PYsK3DnW4Ug.

Garfin, Gregg, Guido Franco, Hilda Blanco, Andrew Comrie, Patrick Gonzalez, Thomas Piechota, Rebecca Smyth, and Reagan Waskom. "Southwest." In *Climate Change Impacts in the U.S.: The Third National Climate Assessment*, edited by Jerry M. Melillo, Terese Richmond, and Gary W. Yohe, 462–86. Washington, DC: U.S. Global Change Research Program, 2014.

Gerke, Sarah Ruth. "A History of National Park Service Interpretation at Grand Canyon National Park." PhD diss., Arizona State University, November 2010. Grand Canyon Museum Collections. https://repository. asu.edu/attachments/56079/content/Gerke_asu_0010E_10100.pdf.

Glaser, Christine, Chuck Romaniello, and Karyn Moskowitz. *Costs and Consequences: The Real Price of Livestock Grazing on America's Public Lands*. Tucson: Center for Biological Diversity, 2015.

Glionna, John M. "Protesters in Utah Drive ATVs onto Federal Land—but Find No Showdown." *Los Angeles Times,* May 10, 2014.

Goad, Ben, and Darrell Santschi, "Baca Bill Aims to Swat Bothersome Fly." *Press-Enterprise,* March 21, 2011.

Goad, Jessica, and Tom Kenworthy. "State Efforts to 'Reclaim' Our Public Lands." Center for American Progress. March 11, 2013. americanprogress.org /issues/green/reports/2013/03/11/56103/state-efforts-to-reclaim-our-public-lands/.

Goldman, Gretchen, Michael Halpern, Deborah Bailin, Arnold Olali, Charise Johnson, and Tim Donaghy. *Progress and Problems: Government Scientists Report on Scientific Integrity at Four Agencies.* Union of Concerned Scientists. October 2015.

Goodwin, Kim, and Dave Burch. "Watch out for Saltcedar." Department of Publications, Montana State University. 2007. store.msuextension.org /publications/AgandNaturalResources/EB0180.pdf.

Goolrick, Faye. *Grand Canyon–Parashant National Monument Long-Range Interpretive Plan.* National Park Service/Pond, August 2012.

Gonzalez, Patrick. "State of Science: Climate Change Impacts and Carbon in U.S. National Parks." *Park Science* 28, no. 2 (2011): 10.

Graves, Lisa. "ALEC Exposed: The Koch Connection." *Nation,* July 12, 2011.

Guo, Jeff. "Wyoming Doesn't Want You to Know How Much Cow Poop Is in Its Water." *Washington Post,* May 20, 2015.

Hamburger, Tom, Joby Warrick, and Chris Mooney. "This Conservative Group Is Tired of Being Accused of Climate Denial—and Is Fighting Back." *Washington Post,* April 5, 2015.

Hannity, Sean. "Ranch Standoff." *Fox News* video. April 9, 2014. youtube.com /watch?v=YtpUpMdigWk.

Hardeen, George. "Black Mesa Railroad Hires First Navajo Woman Operator." *Navajo-Hopi Observer,* July 7, 2011.

Hart Research Associates and North Star Opinion Research. "Strong Bipartisan Support for National Parks—Findings from a National Survey Conducted on Behalf of the National Parks Conservation Association and National Park Hospitality Association." Washington, DC: July 2012.

Hawkins, John. "The 40 Best Political Quotes for 2012 (Fourth Annual)." *Right WingNews,*December13,2012.rightwingnews.com/quotes/the-40-best-political-quotes-for-2012-fourth-annual/#ixzz43dv8B7SK.

Hayes, Chris. "FOX News and GOP Embrace Racist Rancher Cliven Bundy." YouTube video. Posted by Reich-Wing Watch. April 15, 2014. youtube.com /watch?v=PfKZgPlKloE.

Healy, Jack, and Kirk Johnson. "The Larger, but Quieter Than Bundy, Push to Take Over Federal Land." *New York Times,* January 10, 2016.

Hein, Jayni. *Harmonizing Preservation and Production: How Modernizing the Department of Interior's Fiscal Terms for Oil, Gas, and Coal Leases Can Ensure a Fair Return to the American Public.* New York: Institute for Policy Integrity, 2015.

Hellmann, Jessica, James Byers, Britta Bierwagen, and Jeffrey Dukes. "Five Potential Consequences of Climate Change for Invasive Species." *Conservation Biology* 22 (2008): 534–43.

Holmes, Oliver Wendell, Jr. *The Common Law.* Boston: Little, Brown, 1881.

H.R. Res. 5. "Treatment of Conveyances of Federal Land." 115th Cong., 1st sess. January 2, 2017 (q)(1), 35.

Hurlbut, David, Scott Haase, Gregory Brinkman, Kip Funk, Rachel Gelman, Eric Lantz, Christina Larney, David Peterson, and Christopher Worley. "Navajo Generating Station and Air Visibility Regulations: Alternatives and Impacts." Golden, CO: National Renewable Energy Laboratory, March 15, 2012.

IMS Health. "Tamarix Genus Distribution." *Pollen.com.* pollenlibrary.com /map.aspx?map = Tamarix.png.

Jenkins, Kevin. "Bundy: Showdown with Feds a Spiritual Battle." *Spectrum,* August 12, 2014. bit.ly/1pRx8nq.

Jenkins, Peter T. "The Failed Regulatory System for Animal Imports into the U.S.—and How to Fix it." In *Managing Vertebrate Invasive Species: Proceedings of an International Symposium,* edited by Gary William Witmer, William C. Pitt, and Kathleen A. Fagerstone, 85–88. Fort Collins, CO: U.S. Department of Agriculture, Animal Plant Health Inspection Service, National Wildlife Research Center, 2007.

———. "Paying for Protection from Invasive Species." *Issues in Science and Technology* 19, no. 1 (2002).

Jepson, Michael, Tami Clabough, Christopher Caudill, and Russell Qualls. "An Evaluation of Temperature and Precipitation Data for Parks of the Mojave Desert Network." National Park Service, 2016.

Johnson, Marshall. "Navajo Nation Must Move Away from Coal Mining." *Arizona Republic,* June 22, 2010.

Kelly, Erin. "Congress Thwarts Plan to Curtail Grand Canyon Aircraft Noise." *USA Today,* June 30, 2012.

KLAS-8 News. "FBI Investigating Supporters in Bundy Ranch Standoff." YouTube video. May 8, 2014. youtube.com/watch?v=cohNcKlSlcQ.

Kling, Julie. "Preserving Yellowstone: Park Visitation Strains Sustainability Efforts." *Planet Jackson Hole*, November 18, 2014.

Knowlton, Jane. "Home on the Range: What Type of Livestock Graze on National Forests and Grasslands?" U.S. Department of Agriculture (blog). February 3, 2012. blogs.usda.gov/2012/02/03/home-on-the-range--what-type-of-livestock-graze-on-national-forests-and-grasslands/.

Koch Industries. "Life Made Better. It's What We Do." kochind.com /whatwedo/.

———. "Matador Cattle Company." kochind.com/matador/.

Kurzman, Charles, and David Schanzer. *Law Enforcement Assessment of the Violent Extremism Threat*. Durham, NC: Triangle Center on Terrorism and Homeland Security, 2015.

Lange, K. I. "The Jaguar in Arizona." *Transactions of the Kansas Academy of Science* 63 (1960): 96–101.

Leblanc, Steven A. *Prehistoric Warfare in the American Southwest*. Salt Lake City: University of Utah Press, 1999.

Legiscan. "Bill Text: AZ SB 1322, 2012, Fiftieth Legislature 2nd Regular." May 14, 2012. legiscan.com/AZ/text/SB1332/id/553467.

Lenz, Ryan, and Mark Potok. *War in the West: The Bundy Ranch Standoff and the American Radical Right*. Montgomery, AL: Southern Poverty Law Center, 2014.

Leshy, John, and Molly Mcusic. "Where's the Beef? Facilitating Voluntary Retirement of Federal Lands from Livestock Grazing." *New York University Environmental Law Journal* 17, no. 1 (2008): 368–97.

Lessig, Lawrence. *Republic, Lost: How Money Corrupts Congress—and a Plan to Stop It*. New York: Twelve, 2011.

Lieb, Klaus, and Armin Scheurich. "Contact between Doctors and the Pharmaceutical Industry, Their Perceptions, and the Effects on Prescribing Habits." *PLOS One*, October 16, 2014. http://journals.plos.org/plosone /article?id=10.1371/journal.pone.0110130.

Lipton, Eric. "Drillers in Utah Have a Friend in a U.S. Land Agency." *New York Times*, July 27, 2012.

Loomis, Laura. "Congressional Testimony on H.R. 4622, the 'Gateway Communities Cooperation Act of 2002.'" Subcommittee on National Parks, Recreation and Public Lands, Committee on Resources. U.S. House of Representatives. May 7, 2002.

Lustgarten, Abrahm. "End of the Miracle Machines: Inside the Power Plant Fueling America's Drought." ProPublica.com. June 16, 2015. projects.propublica.org/killing-the-colorado/story/navajo-generating-station-colorado-river-drought.

MacEachern, Alan. "Who Had 'America's Best Idea'?" *Niche-Canada.org,* October 23, 2011. niche-canada.org/2011/10/23/who-had-americas-best-idea/.

MapLight. "U.S. Congress, Find Contributions, Ryan K. Zinke." maplight .org/us-congress/contributions.

Markoff, John. "William Hewlett Dies at 87; a Pioneer of Silicon Valley." *New York Times,* January 13, 2001.

Mayer, Jane. *Dark Money: The Hidden History of the Billionaires behind the Rise of the Radical Right.* New York: Doubleday, 2016.

Maza, Cristina. "Secretary Jewell: US Fracking Inspectors Are 'Under-Resourced.'" *Christian Science Monitor,* September 16, 2015.

McCain, John. "Statement by Senator McCain on Grand Canyon Overflights Amendment Passing U.S. Senate." Press release. March 14, 2012. mccain. senate.gov/public/index.cfm/press-releases?ID=124B870F-E5FB AD21–4E35–1A32FAC85E32.

————. *Worth the Fighting For.* New York: Random House: Floris Books, 2002.

McGivney, Annette. "Grand Canyon-Parashant National Monument." *Backpacker,* December 2000.

McGlone, Christopher, Carolyn Hull Sieg, and Thomas Kolb. "Invasion Resistance and Persistence: Established Plants Win, Even with Disturbance and High Propagule Pressure." *Biological Invasions,* June 26, 2010. fs. fed.us/rm/pubs_other/rmrs_2010_mcglone_c002.pdf.

McGlone, Christopher, Judith Springer, and Daniel Laughlin. "Can Pine Forest Restoration Promote a Diverse and Abundant Understory and Simultaneously Resist Nonnative Invasion?" *Forest Ecology and Management* 258 (2009): 2638–46.

McGlone, Christopher, Michael T. Stoddard, Judith D. Springer, Mark L. Daniel, Peter Z. Fulé, and W. Wallace Covington. "Nonnative Species Influence Vegetative Response to Ecological Restoration: Two Forests with Divergent Restoration Outcomes." *Forest Ecology and Management* 285 (2012): 195–203.

Metz, David, and Lori Weigel. *The Language of Conservation, 2013.* Los Angeles: Fairbank, Maslin, Maullin, Metz and Associates, Public Opinion Strategies, 2013.

Moore, Ken, Bob Davis, and Timothy Duck. "Mt. Trumbull Ponderosa Pine Ecosystem Restoration Project." *U.S. Department of Agriculture Forest Service Proceedings.* Washington, DC, 2003.

Moriarty, Thomas. "Sugar Pine Mine Co-owner: 'Please Stop Calling the BLM and Threatening Their Personnel. '" *Medford Mail Tribune,* April 15, 2015.

Morris, Craig. "Renewables Briefly Covered 78 Percent of German Electricity." *Energytransition.de.* July 28, 2015. energytransition.de/2015/07/renewables-covered-78percent-of-german-electricity/.

Morrison, Robyn. "Land Plan Attracts an Anti-grazing Gorilla." *High Country News,* August 5, 2002.

Muir, John. "The Grand Cañon of the Colorado." *Century Illustrated Monthly Magazine* 65 (November 1902–April 1903): 108.

Murphy, Andrea. "Top 20 Largest Private Companies of 2015." *Forbes,* October 28, 2015.

Nash, Steve. "Oil and Water, Economics and Ecology in the Gulf of Mexico." *BioScience* 61, no. 4 (2011): 259–63.

———. "Twilight of the Glaciers." *New York Times,* July 29, 2011.

———. *Virginia Climate Fever: How Climate Change Will Transform Our Cities, Shorelines and Forests.* Charlottesville: University of Virginia Press, 2014.

National Park Service. *Aircraft Management Plan Environmental Assessment.* Grand Canyon National Park, Arizona. 1986.

———. "Air Pollution Impacts: Grand Canyon National Park." nature.nps.gov/air/Permits/aris/grca/impacts.cfm.

———. "Annual Visitation Summary Reports for: 2013–2016." irma.nps.gov/Stats/SSRSReports/National%20Reports/Annual%20Visitation%20Summary%20Report%20(1979%20-%20Last%20Calendar%20Year).

———. "Backcountry Trail Distances." nps.gov/grca/planyourvisit/trail-distances.htm.

———. "California Condors in AZ/UT by Tag #." May 2014. nps.gov/grca/naturescience/upload/CondorChart20140702.pdf.

———. "Clarifying the Definition of 'Substantial Restoration of Natural Quiet' at Grand Canyon National Park, Arizona." *Federal Register* 73, no. 69 (2008): 19248.

———. "Condor Re-introduction and Recovery Plan." nps.gov/grca/learn/nature/condor-re-introduction.htm.

———. "Deferred Maintenance by State and Park." September 2015. nps.gov/subjects/plandesignconstruct/upload/FY-2015-DM-by-State-and-Park.pdf.

———. "Deferred Maintenance by State and by Park," September 30, 2014. nps.gov/subjects/plandesignconstruct/upload/FY14-DM-by-State-and-Park_2015–10–20.pdf.

———. "Endangered, Threatened, and Sensitive Wildlife of Potential Occurrence along the Colorado River in GRCA." nps.gov/grca/learn/nature/upload/threat-endanger.pdf.

———. "Fiscal Year 2015 Budget Justifications and Performance Information." nps.gov/aboutus/upload/FY-2015-Greenbook-Linked.pdf.

———. "Geology Field Notes: Grand Canyon National Park, Arizona." nature. nps.gov/Geology/parks/grca/index.cfm/   azcentral.com/news/election/mccain/articles/2007/03/01/20070301mccainbio-chapter9.html.

———. "Glacier National Park, Going-to the-Sun Road General Info." nps. gov/glac/planyourvisit/gtsrinfo.htm.

———. "Grand Canyon Ecosystems: Cool Canyon Facts." nps.gov/grca/learn/education/upload/EcoArticle-Dec2011–12.pdf.

———. "Grand Canyon National Park, Hydrologic Activity." nps.gov/grca/learn/nature/hydrologicactivity.htm.

———. "Grand Canyon National Park, Special Flight Rules Area in the Vicinity of Grand Canyon National Park: Actions to Substantially Restore Natural Quiet." *Draft Environmental Impact Statement DES 10–60* vol. 1. February 2011.

———. "How Do I Travel to the North Rim?" nps.gov/grca/planyourvisit/directions_n_rim.htm.

———. "Journey through Time: Views of the National Parks." nature.nps. gov/views/Sites/PARA/HTML/ET_04_Journey.htm

———. *Land Resources Division Summary of Acreage.* December 31, 2015. nps.gov/bih/learn/upload/Acrebypark15CY.pdf.

———. "Recreation Visitors by Month, Grand Canyon NP." April 2012.

———. *Report to Congress: Report on Effects of Aircraft Overflights on the National Park System.* September 12, 1994.

———. *Revisiting Leopold: Resource Stewardship in the National Parks.* August 26, 2012.

———. "Stats Report Viewer, Grand Canyon National Park, 2014." irma.nps. gov/Stats/SSRSReports/Park%20Specific%20Reports/Traffic%20Counts?Park = GRCA.

———. "Study of Seeps and Springs." nps.gov/grca/learn/nature/seepspringstudy.htm.

———. "Summary of Acreage." December 31, 2015. irma.nps.gov/Stats/Reports/National.

————. "Tonto Trail: Bright Angel Trail to Hermit Trail." *Grand Canyon*. nps. gov/grca/planyourvisit/upload/Tonto-Bright_Angel_to_Hermit.pdf.

————. "Virtual Visitor Center, Views of the National Parks." nature.nps. gov/views/sites/para/HTML/ET_02_VC.htm.

————. "A Whole Day in Glacier for a Sightseer." nps.gov/tripideas /glacier-whole-day-scenic.htm.

National Public Radio. "How a Historical Blunder Helped Create the Water Crisis in the West." June 25, 2015. npr.org/2015/06/25/417430662/how-a-historical-blunder-helped-create-the-water-crisis-in-the-west.

National Research Council, Committee on Haze in National Parks and Wilderness Areas. *Protecting Visibility in National Parks and Wilderness Areas*. Washington, DC: National Academy Press, 1993.

Neubauer, Chuck and Richard T. Cooper. "In Nevada, the Name to Know Is Reid." *Los Angeles Times,* June 23, 2003.

Newmark, William. "A Land-Bridge Island Perspective on Mammalian Extinctions in Eastern North American Parks." *Nature* 325 (1987): 430–32.

Niemelä, Pekka, and William Mattson. "Invasion of North American Forests by European Phytophagous Insects." *BioScience* 46, no. 10 (1996): 741–53.

Obama, Barack. "Remarks of President Barack Obama: State of the Union Address as Delivered." January 13, 2016. https://obamawhitehouse.archives. gov/the-press-office/2016/01/12/remarks-president-barack-obama____ prepared-delivery-state-union-address.

Ong, Sean, Clinton Campbell, Paul Denholm, Robert Margolis, and Garvin Heath. *Land-Use Requirements for Solar Power Plants in the U.S.* National Renewable Energy Laboratory. June 2013.

"Oregon Militia Leader Goes Berserk over Mailed-In Dildos." YouTube video. January 12, 2016. youtube.com/watch?v=ltpJXLiYVBU.

Painter, Richard W. "The Conservative Case for Campaign-Finance Reform." *New York Times,* February 3, 2016.

Papillon Airways. "Book a Tour: Popular Tours." papillon.com.

————. "Papillon Grand Canyon Helicopters: 50th Anniversary." papillon. com/blog/2015/05/50th-anniversary-celebration.

Pearl, Mike. "What Are Armed Militiamen Really Doing in That Oregon Wildlife Refuge?" *VICE.* January 6, 2016. vice.com/read/whats-really-going-on-in-that-occupied-wildlife-sanctuary-in-oregon-281.

Peck, James. "The Controversy over President Clinton's New Designations under the Antiquities Act." *Arizona Attorney,* July 2000, 10–14, 39–43.

Perez-Pena, Richard. "Disney Drops Plan for History Theme Park in Virginia." *Washington Post,* September 29, 1994.

Pew Research Center. "Public Trust in Government, 1958–2014." November 13, 2014. people-press.org/2014/11/13/public-trust-in-government/.

Pimentel, David, Lori Lach, Rodolfo Zuniga, and Doug Morrison. *Environmental and Economic Costs Associated with Non-indigenous Species in the U.S.* Ithaca, NY. Cornell University College of Agriculture and Life Sciences, 1999.

Pimentel, David, Rodolfo Zuniga, Doug Morrison. "Update on the Environmental and Economic Costs Associated with Alien Invasive Species in the U.S." *Ecological Economics* 52 (2004): 273–88.

Plain, Ronald, and Scott Brown. "Annual Cattle Inventory Summary." Farm Marketing, University of Missouri. February 3, 2014. agebb.missouri.edu/mkt/bull2c.htm.

Povilitis, Tony. "Recovering the Jaguar *Panthera onca* in Peripheral Range: A Challenge to Conservation Policy." *Oryx* 49, no. 4 (2014): 626–31.

Powell, James Lawrence. *Dead Pool: Lake Powell, Global Warming, and the Future of Water in the West.* Berkeley: University of California Press, 2008.

Power, Thomas. "Taking Stock of Public Lands Grazing: An Economic Analysis." National Public Lands Grazing Campaign. August 25, 2003. publiclandsranching.org/htmlres/wr_taking_stock.htm.

Power, Thomas, Donovan Power, and Joel Brown. *The Impact of the Loss of Electric Generation at Glen Canyon Dam: A Report Prepared for the Glen Canyon Institute.* May 1, 2015.

Price, Rae. "Lawsuit over Public Lands Possible in Utah." *Western Livestock Journal,* February 7, 2016.

Public Citizen. "Bundler: Elling Halvorson." citizen.org/whitehouseforsale/bundler.cfm?Bundler = 17271.

Public Employees for Environmental Responsibility. "BLM Ducks Complaint about Suppressing Livestock Damage: Landscape Assessments in Limbo after Scientists Told to Ignore Livestock Impacts." Press release. November 29, 2012.

———. "Employee Violence Report: Rising Tide of Violence, 2015." peer.org/campaigns/whistleblowers-scientists/violence-vs.-employees/rising-tide-of-violence.html.

Ralston, Jon. "Letter from Nevada: The Not-So-Jolly Rancher—How Federal Officials Botched the Bundy Cattle Roundup." *Politico Magazine,* April 28, 2014. politico.com/magazine/story/2014/04/the-not-so-jolly-rancher-106117.

Rampton, Sophie. "Stop the BLM from Restricting Our Land." change.org/p /utah-state-senate-u-s-house-of-representatives-utsgrmp-blm-gov-stop-the-blm-from-restricting-our-land.

Rasby, Rick. "Determining How Much Forage a Beef Cow Consumes Each Day." *UNL Beef.* University of Nebraska–Lincoln. April 2013. beef.unl.edu /cattleproduction/forageconsumed-day.

————. "How Much Water Do Cows Drink Per Day?" *UNL Beef.* University of Nebraska–Lincoln. July 18, 2012. beef.unl.edu/amountwatercowsdrink.

Rasmussen, D. Irvin. "Biotic Communities of Kaibab Plateau, Arizona." *Ecological Monographs* 11, no. 3 (1941): 229–75.

Ratner, Jonathan. "National Park Service Abandons Its Mission." *Western Watersheds Project Messenger,* Fall 2016, 12–13.

"Regulatory Capture." *Investopedia.* investopedia.com/terms/r/regulatory-capture.asp.

Rehfeldt, Gerald E., Nicholas L. Crookston, Marcus Warwell and Jeffrey Evans. "Empirical Analyses of Plant-Climate Relationships for the Western U.S." *International Journal of Plant Sciences* 167, no. 6 (2006): 1123–50.

Reisner, Michael, James Grace, David Pyke, and Paul Doescher. "Conditions Favouring *Bromus tectorum* Dominance of Endangered Sagebrush Steppe Ecosystems." *Journal of Applied Ecology* 50, no. 4 (2013): 1039–49.

Rentz, Catherine, and Rick Young, producers. "Frontline: Flying Cheap." Public Broadcast System. February 9, 2010. pbs.org/wgbh/pages/frontline /flyingcheap/etc/script.html.

Republican National Committee. "America's Natural Resources: Agriculture, Energy, and the Environment." gop.com/platform/americas-natural-resources/.

Research and Polling. "Wolf Recovery Survey: Arizona." June 2008, 1–4.

Reuters. "Thousands Protest Disney History Theme Park Plans." *Los Angeles Times,* September 18, 1994.

Rideout, Bruce A., Ilse Stalis, Rebecca Papendick, Allan Pessier, Birgit Puschner, Myra E. Finkelstein, and Donald R. Smith. "Patterns of Mortality in Free-Ranging California Condors (Gymnogyps californianus)." *Journal of Wildlife Diseases* 48, no. 1 (2012): 95–112.

Ripple, William, and Robert Beschta. "Linking a Cougar Decline, Trophic Cascade, and Catastrophic Regime Shift in Zion National Park." *Biological Conservation* 133, no. 4 (2006): 397–408.

————. "Linking Wolves and Plants: Aldo Leopold on Trophic Cascades." *BioScience* 55, no. 7 (2005): 613–21.

————. "Wolves and the Ecology of Fear: Can Predation Risk Structure Ecosystems?" *BioScience* 54, no. 8 (2004): 755–66.

Rogers, Keith. "Nevada Cattlemen's Association Gives Statement on Rancher Bundy." *Las Vegas Review Journal,* April 21, 2014. reviewjournal.com/news /bundy-blm/nevada-cattlemen-s-association-gives-statement-rancher-bundy.

Romboy, Dennis. "Herbert Signs Bill Demanding Feds Cede Public Lands to Utah." *Deseret News,* March 23, 2012.

————. "Jury Convicts San Juan County Commissioner for Illegal ATV Protest Ride." *Deseret News,* May 1, 2015.

Roosevelt, Theodore. *A Compilation of the Messages and Speeches of Theodore Roosevelt, U.S. President, 1901–1905.* Vol. 1. Washington, DC: U.S. Bureau of Literature and the Arts, 1906.

Rushlo, Michelle. "Monument Plan Draws Cheers, Jeers." *Las Vegas Review-Sun,* January 11, 2000.

Salazar, Adan. "Report: Desert Tortoise Greatly Benefits from Bundy's Cattle Grazing." *Infowars.com.* April 15, 2014. infowars.com/report-desert-tortoise-greatly-benefits-from-bundys-cattle-grazing/.

Skaggs, Jackie. "Creation of Grand Teton National Park (a Thumbnail History)." National Park Service, January 2000. nps.gov/grte/planyourvisit /upload/creation.pdf.

Snow, Nick. "House Approves Bill to Reform the Endangered Species Act." *Oil and Gas Journal,* August 18, 2014. ogj.com/articles/print/volume-112 /issue-8b/general-interest/house-approves-bill-to-reform-the-endangered-species-act.html.

Social Research Laboratory, Northern Arizona University. *Grand Canyon Wolf Recovery, Spring 2005 Grand Canyon State Poll.*

Stoner, Anne M.K., Katharine Hayhoe, Xiaohui Yang, and Donald J. Wuebbles. "An Asynchronous Regional Regression Model for Statistical Downscaling of Daily Climate Variables." *International Journal of Climatology* 33, no. 11 (2012): 2473–94.

Sullivan, Margaret. "The Search for Local Investigative Reporting's Future." *New York Times,* December 5, 2015.

Thomas, Catherine, Christopher Huber, and Lynne Koontz. *2014 National Park Visitor Spending Effects: Economic Contributions to Local Communities, States, and the Nation.* National Park Service, April 2015.

Town of Tusayan. "Town Leadership." tusayan-az.gov/leadership/.

Salt Lake Tribune. "If Utah Can Help Mexican Wolves Recover, We Should Let Them In." December 5, 2015.

————. "Where's the Money?" July 7, 2013.

Salt River Project. "Facts about SRP." srpnet.com/about/facts.aspx.

————. "Navajo Generating Station." www.srpnet.com/about/stations /navajo.aspx.

Shedlowski, Kirby-Lynn. "Grand Canyon to Replace Portion of Trans-Canyon Pipeline at Phantom Ranch." Press release. January 23, 2015. nps.gov /grca/learn/news/replace-portion-tcp.htm.

Silverman, Amy. "The Gift That Keeps on Giving Access." *Phoenix New Times,* November 25, 1999.

Snider, Annie. "Colorado River: Water Users Scramble as Drought Foretells Scary Future." *Greenwire,* December 24, 2014.

Sonner, Scott, and Martin Griffith. "Tension Growing between Ranchers, Mustang Backers." *Salt Lake Tribune,* April 6, 2014.

Sorensen, Christopher. "Ponderosa Pine Understory Response to Short-Term Grazing Exclusion (Arizona)." *Ecological Restoration* 28, no. 2 (2010): 124–26.

Stein, Sam. "Koch Brothers Group Wipes Cliven Bundy Support from Social Media Accounts." *Huffington Post,* April 25, 2014. huffingtonpost.com /2014/04/25/koch-brothers-cliven-bundy_n_5213481.html.

Stokowski, Patricia, and Christopher LaPointe. *Environmental and Social Effects of ATVs and ORVs: An Annotated Bibliography and Research Assessment.* School of Natural Resources, University of Vermont. November 20, 2000.

"SW Climate Outlook." *Climate Assessment for the Southwest.* University of Arizona. July 17, 2014. climas.arizona.edu/swco/jul-2014/southwest-climate-outlook-july-2014.

Synapse Energy Economics. "Renewable Energy Alternatives to NGS." August 23, 2013. Cambridge, MA.

Taylor, Phil. "Utah Official Convicted in Recapture Canyon Protest Ride." *E&E News.* May 4, 2015. eenews.net/stories/1060017929.

Teachout, Zephyr. *Corruption in America.* Cambridge, MA: Harvard University Press, 2014.

Thompson, Dennis. *Ethics in Congress: from Individual to Institutional Corruption.* Washington, DC: Brookings Institution Press, 1995.

Tilman, David, Peter Reich, and Johannes Knops. "Biodiversity and Ecosystem Stability in a Decade-Long Grassland Experiment." *Nature* 441 (2006): 629–32.

Turkewitz, Julie. "Drought Pushes Nevada Ranchers to Take on Washington." *New York Times,* July 2, 2015.

United States of America v. Cliven Bundy. U.S. District Court for the District of Nevada. CV-S-98–531-JBR (RJJ). November 3, 1998.

United States of America v. Cliven Bundy. U.S. District Court for the District of Nevada. CV-0804-LDG-GWF. July 9, 2013.

USA Today. "Militia Leader Meets Oregon Sheriff, Refuses to Leave." YouTube video. January 9, 2016. youtube.com/watch?v=TebwkuIDw6w.

U.S. Bureau of Labor Statistics. "CPI Inflation Calculator." www.bls.gov/data /inflation_calculator.htm.

———. "Report 1043: Mass Layoffs in 2012." BLS Reports. September 2013.

U.S. Congress. "Endangered Species Act Amendments of 1978." Sec. 7 Washington, DC: 1978.

———. "Establishment of a Peace Garden and Minimum Altitude for Aircraft Flying over National Park System Units: Hearing before the Subcommittee on Public Lands, National Parks, and Forests of the Committee on Energy and Natural Resources." U.S. Senate, 100th Cong., 1st sess. May 7, 1987 (statement of Sen. John McCain).

———. "Report, House of Representatives, 2d Session, Federal Aviation Authorization Act of 1996, Conference Report to accompany H.R. 3539." September 26, 1996.

U.S. Congress, Office of Technology Assessment. *Harmful Non-indigenous Species in the United States.* Washington, DC: U.S. Government Printing Office, 1993.

U.S. Department of Agriculture, Animal and Plant Health Inspection Service, Wildlife Services Program. "2014 Program Data Report G: Animals Dispersed/Killed or Euthanized/Freed." aphis.usda.gov/aphis/ourfocus/wild lifedamage/sa_reports/sa_pdrs/sa_2014/ct_pdr_g.

———. "Forest Service Kaibab National Forest Returns Easement Application to Town of Tusayan." Press release. March 4, 2016. fs.usda.gov/detail /kaibab/news-events/?cid=FSEPRD493806.

U.S. Department of Commerce, Bureau of Economic Analysis. "U.S. International Trade in Goods and Services." December 2015. bea.gov/newsreleases /international/trade/tradnewsrelease.htm.

U.S. Department of Homeland Security, Office of Intelligence and Analysis. "Domestic Violent Extremists Pose Increased Threat to Government Officials and Law Enforcement." July 22, 2014.

U.S. Department of the Interior. "Secretary Salazar Announces Decision to Withdraw Public Lands near Grand Canyon from New Mining Claims."

Press release. January 9, 2012. doi.gov/news/pressreleases/Secretary-Salazar-Announces-Decision-to-Withdraw-Public-Lands-near-Grand-Canyon-from-New-Mining-Claims.

U.S. Federal Aviation Administration. *Calendar Year 2014 Passenger Boardings at Commercial Service Airports.* September 2015. faa.gov/airports/planning_capacity/passenger_allcargo_stats/passenger/media/cy14-commercial-service-enplanements.pdf.

U.S. Fish and Wildlife Service. *Draft Mexican Wolf Revised Recovery Plan,* September 16, 2011.

———. "Environmental Impact Statement for the Proposed Revision to the Nonessential Experimental Population of the Mexican Wolf." Draft. July 16, 2014.

———. "Investigation Complete for Wolf Killed in Utah." July 9, 2015. *News and Releases.* fws.gov/mountain-prairie/pressrel/2015/07092015_Investigation-Complete-for-Wolf-Killed-in-Utah.php.

———. *Mexican Wolf Blue Range Recovery Area Reintroduction Project Middle Management Team, Estimated Funds Expended by Lead Agencies for Mexican Wolf Recovery and Reintroduction.* December 31, 2013.

———. "Mexican Wolf Recovery Area Statistics." 2015. fws.gov/southwest/es/mexicanwolf/pdf/MW_popcount_web.pdf.

———. *Mexican Wolf Recovery Program 2015 Progress Report.*

———. "Mexican Wolf Recovery Timeline." *Ecological Services.* December 31, 2015. fws.gov/southwest/es/mexicanwolf/chronology.cfm.

———. *Recovery Outline for the Jaguar (Panthera onca).* April 2012.

———. "Service Designates Jaguar Critical Habitat in Arizona and New Mexico." *Arizona Ecological Services.* Press release. March 4, 2014. fws.gov/southwest/es/arizona/Jaguar.htm.

———. "Statistical Data Tables for Fish and Wildlife Service Lands (as of 9/30/2015)." fws.gov/refuges/land/PDF/2015_Annual_Report_of_Lands-DataTables.pdf.

U.S. Fish and Wildlife Service, Mexican Spotted Owl Recovery Team. *Recovery Plan for the Mexican Spotted Owl.* November 2012.

U.S. Fish and Wildlife Service, U.S. Department of Commerce, and U.S. Bureau of the Census. *1996 National Survey of Fishing, Hunting, and Wildlife-Associated Recreation.*

———. *National Survey of Fishing, Hunting, and Wildlife-Associated Recreation.* 2011.

U.S Forest Service. "By the Numbers." 2013. fs.fed.us/about-agency/newsroom/by-the-numbers.

————. *Land Areas of the National Forest System.* January 2012. fs.fed.us/land/staff/lar/LAR2011/LAR2011_Book_A5.pdf.

————. "Response to a Freedom of Information Act Request, #2016-FS-WO-03262F, Stephen Nash."

————. *The Rising Cost of Fire Operations: Effects on the Forest Service's Non-fire Work.* August 4, 2015.

U.S. General Accounting Office. *Invasive Species: Obstacles Hinder Federal Rapid Response to Growing Threat.* Washington, DC: U.S. General Accounting Office, 2001.

U.S. Geological Survey, Colorado Plateau Research Station. *The National Map.* 2016. https://nationalmap.gov/#.

U.S. Government Accountability Office. "Federal Lands: Enhanced Planning Could Assist Agencies in Managing Increased Use of Off-Highway Vehicles." *Report to the Subcommittee on National Parks, Forests and Public Lands, Committee on Natural Resources, House of Representatives.* June 2009.

————. *Federal User Fees: Key Considerations for Designing and Implementing Regulatory Fees.* September 2015. gao.gov/assets/680/672572.pdf.

U.S. Office of Natural Resources Revenue, Department of the Interior. "Reported Revenues, All Land Categories in All States and Offshore Regions for Fiscal Year 2014." statistics.onrr.gov/ReportTool.aspx.

Vanderbilt, Amy, and Ron Moler. "Saving a National Treasure. FHWA and the National Park Service Embark on a Monumental Restoration of Montana's Historic Going-to-the-Sun Road." *Public Roads* 70, no. 3 (2006).

Vincent, Carol, Laura Hanson, Jerome Bjelopera. *Federal Land Ownership: Overview and Data.* Washington, DC: Congressional Research Service, 2017.

Vogel, Kenneth. "The Kochs Put a Price on 2016: $889 million." *Politico,* January 26, 2015.

Vucetich, John, and Michael Paul Nelson. "Conservation, or Curation?," *New York Times,* August 20, 2014, A 21.

Wagner, Dennis. "Macho B: Cover-Up amid Celebrations." *Arizona Republic,* December 10, 2012.

Walters, David, Emma Rosi-Marshall, Theodore Kennedy, W.F. Cross, and Colden Baxter. "Mercury and Selenium Accumulation in the Colorado River Food Web, Grand Canyon, U.S.A." *Environmental Toxicology and Chemistry* 34, no. 10 (2015): 1–10.

Walters, Jeffrey, Scott Derrickson, D. Michael Fry, Susan M. Haig, John M. Marzluff, and Joseph M. Wunderle. *Status of the California Condor and Efforts to Achieve Its Recovery.* American Ornithologists Union. August 2008. www.fws .gov/cno/es/calcondor/PDF_files/AOU-Audubon2008-Report.pdf.

Wild Wilderness. "Background Information on: American Recreation Coalition." wildwilderness.org/docs/94bod.htm.

Wyoming State Legislature. "Original Senate Engrossed File N. SF0012, Enrolled Act No. 61." Sixty-Third Legislature of the State of Wyoming, 2015.

Yen, Hope, and Thomas Peipert. "Four in 10 Higher-Risk Oil and Gas Wells in U.S. Aren't Inspected by Feds." *Washington Post,* June 15, 2014.

INDEX

AARP (American Association of
Retired Persons), 92
Abbey, Robert, 173–174
Agricultural Plant Health Inspection
Service (APHIS), 22–25, 194
agriculture, in the Grand Canyon
area, 2, 28–29
air quality, 2, 95–113; federal regulation
and enforcement, 98–102, 103, 108,
230; at Grand Canyon National
Park, 96–98, 99, 101, 103, 107, 181; the
human health trade-off, 105–108; the
jobs trade-off, 104–105, 109–110, 112,
natural pollution sources, 98; the
Navajo Generating Station, 102–105,
106, 108–112, Map 2; and vehicle
traffic, 181; the view impairment
trade-off, 96–98, 101, 109
air tourism: the FAA's role in overflight
regulation, 182, 184, 187–188, 190, 195,
203, 205, 206; at Grand Canyon
National Park, 69, 90–91, 178–179,
181–185, 187–192, 201–206; and the
Hoffman changes to Park Service
policies, 186; industry lobbying and
political contributions, 191–192,
195–196; industry opposition to Park

Service noise regulation, 185,
189–192, 195; safety problems, 182,
184, 203. See also noise
aircraft noise: commercial jets, 187–188.
See also air tourism; noise
alien invasive species. See invasive
exotic species
American Association of Retired
Persons (AARP), 92
American Lands Council, 172, 236
American Legislative Exchange
Council, 172–173
American Prairie Reserve, 168
American Recreation Coalition, 185
Ancestral Puebloans, 2, 27–29, 65–66,
163, 222
Anderson, Mark, 36
anglers. See hunting and fishing
animal pests, 16, 25, 32. See also grazing
entries; invasive exotic species
animals. See animal pests; threatened/
endangered species protection;
wildlife
antifederalism, 144, 148; the Bundy
standoff (2014) and its aftermath,
160–164, 165, 166; historical
precedents, 164–165; law-

public engagement, 9; on park funding challenges, 81, 87; on park user fees, 91, 92; on Tusayan incorporation and development proposals, 69–70, 71, 76, 212

Uncle John problem. *See* jobs vs. environment debates

unemployment. *See* economic impacts; jobs vs. environment debates

Unkar Delta, 2, Map 2

uranium mining, 210–211

U.S. Air Tour Association, 185, 195. *See also* air tourism

U.S. Congress: attempts to weaken the Endangered Species Act, 55–56, 85; and federal lands relinquishment, 4; House hearings on FAA safety oversight, 204–205; National Park Service budget appropriations, 86, 93; the need for public engagement, 9. *See also* Arizona congressional delegation; political corruption and influence; Republican Party

U.S. government agencies. *See* Bureau of Land Management; Federal Aviation Administration; Fish and Wildlife Service; *other specific agencies by name*

U.S. government lands. *See* public lands *entries; specific locations and administering agencies*

USDA: import inspections, 22–24, 194; Wildlife Services Program, 146

user fees, 90–94, 229

Utah: anti-government views in, 148; and efforts to gain control of federal lands, 170; state and local government opposition to wolf reintroduction, 50–51

Utah State Cattlemen's Association, 166

vandalism, 163, 180

vehicles, in national parks, 179–181

Vermilion Cliffs National Monument, 3, 42, 168, 169, 175, Map 2, Map 8

views and visibility, air quality and, 96–98, 101, 109

violence and threats, 148; the Bundy standoff and its aftermath, 160–164, 165, 166

Virginia, the Disney's America theme park proposal, 75–76, 211

visitor fees, 90–94, 229

Vucetich, John, 37, 49–50, 56, 62

Wade, Bill, 186–187

Walcott, Charles, 46

Walhalla Plateau, 27–29, Map 1

Walt Disney Company, Disney's America theme park proposal, 75–76

Walters, Jeffrey, 41, 42, 44–45, 46

Washington County, Utah: BLM form letter, 148; opposition to wolf reintroduction, 50–52

Washington, George, 165

water: ancient irrigation systems on the Walhalla Plateau, 28; Colorado River diversions and allotments, 103–104, 110–111, 213–214; the fragility of Grand Canyon water sources, 64–65, 66, 71, 78, 81–82, 88–89; hydrological impacts of grazing, 122, 152–154; hydrological impacts of plant pests, 19, 26; the NGS and subsidized water supplies for southern Arizona, 103–104, 110–111; potential hydrological impacts of climate change, 29, 66, 71, 89; potential hydrological impacts of Tusayan development, 66, 67, 70, 71. *See also* Colorado River

water quality: impacts of grazing, 122; impacts of off-road vehicles, 180; mercury pollution, 109; mining contamination at Grand Canyon National Park, 210

Wechsler, Robert, 69

weeds. *See* plant pests

Whiskey Rebellion, 165